SUN TZU
AND THE ART OF MEDIEVAL
JAPANESE WARFARE

Kashima-Shintō-ryū: an *okuden*-level *ch'i* counter-attack; the 'unexpected' used against the 'expected'. (Photo courtesy Phil Jupp, British Kendō Renmei, taken at Kashima-jingu, 1982.)

SUN TZU
AND THE ART OF MEDIEVAL JAPANESE WARFARE

Roald Knutsen

Roald Knutsen
17th November 2006

GLOBAL ORIENTAL

SUN TZU AND THE ART OF MEDIEVAL JAPANESE WARFARE

by Roald Knutsen

First published 2006 by

GLOBAL ORIENTAL Ltd
PO Box 219
Folkestone
Kent CT20 2WP
UK

www.globaloriental.co.uk

© 2006 Roald Knutsen

All rights reserved. No part of this publication may be reproduced or transmitted in any form or by any electronic, mechanical, or other means, now known or hereafter invented, including photocopying and recording, or in any information storage or retrieval system, without prior permission in writing from the Publishers, except for the use of short extracts in criticism.

British Library Cataloguing in Publication Data
A CIP catalogue entry for this book is available from the British Library

ISBN 978-1-905246-00-7

Set in Stone Serif 9.5 on 13pt by Servis Filmsetting Ltd, Manchester
Printed and bound in England by Antony Rowe Ltd, Chippenham, Wilts

Contents

[See plates facing page 100]

Acknowledgements	ix
List of Illustrations	xi
Introduction	1

CHAPTER 1
The Muromachi Warriors and their Approach to Sun Tzu: Some Historical Benchmarks	12
Role of the Deities	18
Bugei Origins	22
Focus on Practical Experience	25

CHAPTER 2
Who Were the Bugeisha?	28
The *Seven Military Classics*	30
Underdetermined Metaphors	33

CHAPTER 3
Ch'i Within the Eishin-ryū	36
The 'Layered' Structure	39
Invisibility	44

CHAPTER 4
Hasegawa Eishin-ryū Structure 46
    1. Ōmori-ryū 46
    2. Musō-Jikiden Eishin-ryū Iai-hiza Forms 48
    3. Okuden – Iai-hiza Forms 48
    4. Okuden – Tachi-waza Forms 49
    Alternative Forms 50

CHAPTER 5
Mao Tse-tung and Unorthodox Tactics 52
    Distraction and Concentration 55

CHAPTER 6
Iai-jutsu Seen as Flexible Warfare 59
    The Chinese Arts of War as Interpreted by the Bushi 61
    'All Warfare Is Based on Deception' 65

CHAPTER 7
Foreknowledge 71
    Shih 73
    Naturalness of Movement 75
    In and *Yō* 83
    Summary 86
    Kamae 87

CHAPTER 8
The Distinction in the Heihō between *Ch'i* and *Cheng* 91
    The Rationale 92

CHAPTER 9
The Influence of the Mountain Religion 95

CHAPTER 10
The Esoteric Principles Contained in *In* and *Yō* 102
    The Griffin Totem and Marishi-ten 105

CONTENTS

CHAPTER 11
Unexpected Attacks Against an Unprepared Enemy 112
   'Threaten in the East; Strike in the West' 114
   Cheng to Ch'i 117
   Ch'i to Ch'i 118
   Diversionary Tactics 119
   Striking from Concealment 121
   Strike Where the Enemy Does Not Expect 123
   Coiling-Smoke Principle Applied 126
   Divide and Rule 128
   Making Use of Cover Prior to a Surprise Attack 129

CHAPTER 12
Fundamental Teachings 132

CHAPTER 13
Warfare and Ritual 135
   The Manuals of the Arts of Warfare 136
   The Developments in Europe 138
   Comparisons 145
   Tengu Iconography 146
   Yagyū Shinkage-ryū Tengu 150
   Taisha-ryū Tengu 151
   Other Tengu Illustrations 153

CHAPTER 14
Conclusions 155
   Examples of the Terminology Used in Heihō 157

CHAPTER 15
'Subtle and Insubstantial . . .' 164

Notes 169
Glossary of Terms 177
Bibliography 184
Index 187

Acknowledgements

It is difficult to adequately acknowledge those many people who contribute to one's lifetime experience. I am deeply grateful to all my teachers within the field of Bujutsu and Budō, past and present, though I hesitate out of deference to name those still with us. However, I will express my heartfelt gratitude to the memory of Arai Shigeo, Kamō Jisaku, Yoshikawa Koichirō, Katō Urasaburō, and to Takami Taizō, all *sensei* who first set my feet on the proper path both of Iai and Kenjutsu. Behind them, standing in the shadows, were a number of other even more senior masters, amongst them Ozawa Takashi Hanshi and Matsumoto Toshio Hanshi. The first roots of this study go directly back to a number of conversations when I was serving in Tripoli, Libya, during the Suez Crisis of 1956 when a brilliant trio of my fellow NCOs in Field Security informally discussed and analysed the beginnings of the Vietnam War and its background of the previous conflict in French Indo-China. I shall never forget those discussions as they first pushed me towards what I later understood to be *heihō*. At the present time, I am also grateful to Mrs Nakano Tomoko and Mrs Naomi Phillips for their several valuable translations, and to Kinyo and Phil Jupp for their input. From the historical perspective, too, I defer to the guidance of Russell Robinson, Dr Kuroki Toshihiro and Dr Benjamin Hazard, though I hasten to point out that any errors and deductions are entirely mine. I am for ever in debt to Ric Bithell for his computer expertise and so carefully reading and correcting my text; to my wife, Patricia, with her long experience of Kendō and Naginata and her generous patience that has allowed me the opportunity to return to Japan many times despite all our family commitments. Lastly, and very importantly, I thank Paul Norbury, my publisher, for his encouragement and support. I am indebted to you all for helping open yet another 'window' on Old Japan.

List of Illustrations

[*See plates facing page 100*]

1. Marume Kurando-no-suke
2. Memorials to members of the Yagyū family
3. Wooden painted statue of Yagyū Munenori
4. Fujiwara Idzumo Ise-no-kami Nobutsuna's four scrolls
5. Shinkage-ryū
6. Letter written by the thirteenth Shogun, Ashikaga Yoshiteru
7. Licence presented to Marume Kurando-no-suke by Ise-no-kami Nobutsuna
8, 9. Illustrations from the Fuzeiken-Taisha-ryū kumitachi scroll
10. Example of tengu in the role of *uchidachi* (attacker)
11. Tengu named Chiraten (-bō)
12. Iai-jutsu example: a sudden evasive retreating move employing a turning draw and thrust
13. Iai-jutsu example: using strong sweeping attacks at two different angles
14. Iai-jutsu example: using a turn towards the enemy to the rear accompanied by a devastating upwards *kesagiri* stroke and final strong *kirioroshi*
15. Iai-jutsu example: feigning retreat to create an opportunity to turn and destroy an unprepared pursuing force
16. A kenjutsu form of the Yagyū Shingan-ryū
17. An unusual in-no-kamae

18. The end of a form; Yagyū Shinkage-ryū
19, 20, 21. Naginata-jutsu and kenjutsu demonstrated by masters at the Kashima Shintō-ryū
22, 23. Kenjutsu of the Yagyū Shinkage-ryū
24. A tengu-tobi leap from a half-kneeling posture
25, 26. Bo-jutsu (uchidachi) opposed by sword
27. Kenjutsu
28. Various symbols mentioned in the text
29. Other cultic symbols that became associated with Marishi-ten (Mārīcī)

Introduction

Just on half-a-century ago, a few months after three-and-a-half years' army service with the Intelligence Corps, mostly on active duty, I started out on my Budō career joining one of the three foremost Judō dōjō in London. I was fortunate enough to find there some Kendō *bogu*, armours, and the encouragement to use them. Within the following four years Kendō began to take root. This is due, in no small part to: increasing contacts with the Japanese community; the guidance of the late H. Russell-Robinson of the Royal Armouries who properly introduced me to the intricacies of Japanese arms, armour and warrior history; the background support of H.E. the Japanese Ambassador at that time and his Military Attaché, both Kendō *yūdansha* of note; a regular and constant correspondence with a Californian master, Benjamin H. Hazard, seventh *dan*; and, finally, the increasing interest of four leading members of the All Japan Kendō Renmei. My earlier teenage enthusiasm inclining towards 'Things Japanese' was now firmly channelled towards the proper study of the background to these warrior arts and, together with my wife, Patricia, I began to identify and separate the traditional from the modern developments of the twentieth century.

Around 1962, almost as soon as I embarked on the study of Iai proper I came under the guidance of Takami Taizō *sensei*, pre-war *go-dan* in both Kendō and Iai. I bought a copy of Samuel Griffith's translation of *Sun Tzu, The Art of War (=Ping Fa)*, a book that has had a huge influence on the whole of my subsequent *bugei* career. I can honestly say that this work written two-and-a-half millennia ago, has been the key to understanding the arts of Bujutsu and the concepts that shaped them. One of General Griffith's comments that I read at that time greatly intrigued me since it echoed my specialist experience in Field Security in the army. After surveying the historical development of interest in Sun Tzu from earliest times through to the Second World War in Japan, he pointed out that:

> It appears that in spite of devoted study the Japanese understanding of Sun Tzu was no better than superficial. In the most profound sense, they knew neither their enemies nor themselves . . . and they had forgotten, too, the wise words of Mencius: 'And so it is certain that a small country cannot contend with a great, that few cannot contend with many; that the weak cannot contend with the strong.'

This comment remained at the back of my mind through the ensuing years of practice and the good fortune, over the last thirty of them, to be exposed to classical Bujutsu under a number of really excellent teachers both here and in Japan. Increasingly, throughout that period, I resolved to test Griffith's analysis and came to the conclusion that whilst it was absolutely true of the Japanese military and political thinking that led up to the tragedy of military expansion through the 1920s and 1930s and one of the root causes of their eventual defeat in 1945, it was the misunderstanding of the principles outlined by Sun Tzu and his grandson (?), Sun Pin, that led to the final disaster. But how were these principles actually studied and, therefore, *who* did understand them and when?

To my mind, part of the answer lay right in front of me in the structure and content of the classical arts of the bugei. I make no claim to insight into modern military tactics and strategy but I have some understanding of how the masters of *heihō* tried to interpret, illustrate and transmit the principles within the structures of many diverse systems during that turbulent period of Japanese history, the Muromachi period (roughly from 1350 to 1575). It was in these two centuries that a number of gifted warriors developed the arts of warfare, as they saw them, to heights of formidable effectiveness little understood in the West. In these pages no attempt has been made to examine the strengths or weaknesses of the medieval Japanese command structures, this is a matter for war historians, but to focus instead on certain basic strategies incorporated by the master swordsmen of the Muromachi period and expressed in their often unique systems, then kept very secret by them or developed on the field of battle when some of these same swordsmen were employed as strategists by their clan lords.

Sun Tzu made two telling statements in the *Ping Fa*; the first was that: 'War is a matter of vital importance to the State; the province of life or death; the road to survival or ruin.' The second was: 'All warfare is based on deception.' The first is, possibly, stating the obvious, but it is the second statement that is at the centre of the medieval Japanese understanding of Sun Tzu and the corpus of ancient Chinese works known collectively as *The Seven Military Classics*.

It has not been my intention here to cite specific examples from named traditions or to closely identify those *ko-ryū*; rather the objective is to throw more light on the deadly serious understanding that many *kenshi* possessed and included in their systems, often never divulging these secret principles to anyone outside their immediate circle during their lifetimes. Such masters of the bugei considered these matters so important they transmitted only by oral teaching to those of their students they considered worthy of becoming custodians of such developed knowledge. The fact that such an attitude remained intact throughout the medieval period within many of these ryū, and continues to the present day, does not mean that they should now be opened up and exposed to anyone. To reach the stage of understanding these matters demands constant practice over many years, even decades. Their integrity is of the greatest importance. The late Donn Draeger once was requested to introduce a small group of high-ranking Kendō masters to a ko-ryū dōjō practising in a mountain village northeast of Tokyo. After watching some truly unique Bujutsu that originated in the Sengoku period, one of these masters commented that it was a pity that only three of the ryū masters were present and they were quite elderly; were they not afraid that this tradition would pass away with them and be lost? The reply was that this particular ryū had been this size since the seventeenth century. If it was lost then so be it! At least it would remain intact right up to that point. These are precisely the same sentiments shared by my wife and myself. The integrity of these traditions is all-important, and so is respect for both the teachings *and* the masters who have so carefully handed them down through many generations. Reigi is always at the foundation of true understanding and should never be forgotten; casual lip-service to this is not enough.

Quite possibly this attitude is an archaic anachronism in the twenty-first century, but then, we are discussing in these pages military secrets as they were viewed in the Muromachi age. Then they were considered 'mouth to ear' matters; why change this now? Where I have made specific references to any ko-ryū system it will mostly be to the teachings within the Hasegawa Eishin-ryū, but even then I shall avoid too much actual detail. Those of my readers who may wish to enquire further will have to be prepared to undergo the long years of apprenticeship that all students have experienced. There is no easy method and I am sorry to say that in this area relatively short-cut academic pathways do not and never have existed.

While most of the principles of warfare were outlined by Chinese masters like Sun Tzu, once they reached and were developed in Japan over a thousand years later they were refined and 'Japanicized' as was everything else imported into the archipelago. A number of esoteric influences were soon introduced and in themselves these are of great interest, not to say, significance. Certainly, they cannot be ignored. What is important is to distinguish between the developments and interpretations of the Muromachi period when warfare was both endemic and often on a large scale, and to ignore the long period of change that altered Bujutsu to Budō that started after the battle of Sekigahara in 1600. The two-hundred-and-sixty or so years of comparative stability presided over by the Tokugawa *bakufu* permitted other influences to soften and change a large proportion of martial theory, often to a major extent, and divert it significantly away from that of the turbulent times that were the final crucible of fire for the bugei. Some of these misinterpretations and weaknesses entered military thinking in the late-Edo period and early modern times, particularly it seems in the Imperial Japanese Army though not to the same extent in the Imperial Japanese Navy; this is not my area of focus and in these pages we shall confine ourselves to examining aspects of the warrior rationale that preceded the Tokugawa shogunate.

When we approach the arts of the bugei objectively, we find that many of the swordmasters active in the mid- and late-Muromachi age, perhaps spurred on by the unstable political environment of their period, significantly focused their thinking on the writings of Sun Tzu and used his *Ping Fa* as the springboard to develop their personal interpretation and understanding. These able warriors, known amongst the samurai as *kenshi*, were individuals and remained individuals despite acquiring great skill; it was natural for them to rationalize the arts of war in terms of contests of one 'general' versus an enemy 'general'. This viewpoint was already emphasized by the teachings of the ancient Chinese military philosophers which were often couched in terms of a discourse between the strategist and his ruler. From the Muromachi period at least, Bujutsu, and later Budō, has taken the form of instruction given on a one-to-one basis between master and pupil.

It is hardly surprising that when we come to examine the techniques of the various systems we find that they tend to concentrate on warfare based on deception. Around two-thousand-five-hundred years ago, Sun Tzu and others who succeeded him discoursed on what they meant by deception; in modern times a brilliant war leader of the calibre of Mao Tse-tung, a master of guerrilla warfare, rationalized deception in the field in such a way that his unlettered commanders, often drawn from the ranks of the peasantry, were able to take on and gradually drain and destroy the might of the Imperial Japanese Army. One of the determining principles for military success is intelligence; the more that can be found out about a potential enemy and, conversely, the more that same type of information can be denied him, the better. The *Four Poisons* of the bugei arts are defined as 'fear', 'doubt', 'indecision' and 'surprise' and it is interesting that apart from 'fear' the other three have everything to do with what Griffith describes as 'the complementary arts of simulation and dissimulation'. Add in the ingredient of 'fear' and the problems for a general become formidable indeed.

Intelligence and counter-intelligence have everything to do with 'shaping' the enemy and not being 'shaped' by him. Intelligence gathered by the use of agents, and particularly by those men who are experts in the bugei and from whom little can be hidden, permit the wise general to attack the mind of his opponent. Throughout those ko-ryū that have descended to us more or less intact from the Muromachi period we find the closest attention devoted to increasing the students' awareness of the need to 'shape' their opponents. In the various systems of Iai-jutsu, where the opponent is not actually present, only visualized, this 'shaping' is paradoxically absolutely essential. This is because, of all the warrior arts that comprise the classical Bujutsu traditions, Iai-jutsu, in particular, tackles the problems of the 'normal', or 'orthodox', actions at a very close interval against the enemy and makes use of the 'extraordinary', or 'unorthodox'. This is not to imply that the other major systems do not do this; it is just that in the Battō, or 'Drawing Sword' arts of the Muromachi, we find the 'unorthodox' applications addressed directly.

Normally generals take up positions and prepare their tactics according to their understanding of their situation and that of the opposing side. This was termed the use of *cheng* or direct force from ancient Chinese times. When faced with an enemy of, say, much greater force, such *cheng* dispositions might be impossible, even foolhardy, and the employment of 'indirect' or *ch'i* far more productive. The *ch'i*, or unorthodox force, can be used to probe, encircle, outflank the enemy, to attack his lines of communication, to force him to deploy troops in areas that he feels are vital to his strategy, to attack other places which he does not yet defend, to divide him, to mislead, confuse and even entrap him. There are endless possibilities and equally endless permutations in the combining of the *cheng* and *ch'i*. The principles are all contained in the Iai-jutsu teachings and in the case of the oldest of the classical ko-ryū, dating without doubt from the middle of the Muromachi period, are embedded within the structures devised by the talented kenshi of that era.

Analysis of the forms within these classical traditions, and especially those that compose the upper levels, known as the *okuden* or secret transmissions, reveals an amazing multiplicity of simple but brilliant interpretations of the ancient Chinese precepts. These are truly matters of the greatest concern to the State, whatever its size – from the local clan to the rulers of domains stretching across several provinces. Above all, they were preserved as the most covert of secrets. In the words of one of Sun Tzu's distinguished commentators already quoted above: 'These are "mouth to ear" matters.' The later Chinese scholar, Chia Lin, is aptly cited by General Griffith: 'An army without secret agents is exactly like a man without eyes or ears.'

We shall also examine something of the deep influences exerted on the development of this warrior understanding of the Chinese classics through the influence of the female warrior deity, Marishiten, that entered Japan around the eighth century from the Asian mainland, and other factors that are even earlier. The Japanese swordsmen, or bugei-sha, drew on very old and long-established beliefs in their quest to rationalize and understand the principles of warfare. They did not have the advantage, if advantage it is, of modern explanations of such things as 'intuitive comprehension' and so on to explain their 'creative leaps' in the application of the weaponed arts and generalship; they did have an unquestioning conviction that such intuitive inspiration reached them through some 'outside' agency. This they interpreted as 'a gift from the deities'; or divine revelation.

The background to these beliefs in the guidance bestowed by divinities or the metaphysical world can, I am quite sure, be detected as early as the mid-Kofun period of the fourth and fifth centuries and quite possibly entered Japan in the hands of a small group of warrior-shaman at the end of the Yayoi period in the second or third century CE, soon to be absorbed into the pervasive background of the indigenous folk beliefs that remain so strong in rural Japan. By the sixth or seventh century these particular warrior-shaman seem to have become the proto-yamabushi who lived in small groups in certain mountain fastnesses and, it is said, 'protected' the many Taoist recluses who sought these wild places to gain metaphysical powers through a variety of magical and semi-religious rituals. This was fertile ground indeed when, at the beginning of the ninth century, esoteric Buddhism was introduced with its huge immediate appeal to many sections of society. Throughout the ensuing Heian period, the ninth to the twelfth centuries, these esoteric influences of both Tendai and Shingon Buddhism and the secretive Shugendō sects, began to find a response in the growing martial philosophies that formed the basis of the bugei, and deepened rapidly as the political controls of the Kamakura, Hōjō and early Muromachi administrations disintegrated into periods of what might be seen as near anarchy.

From the mid-fifteenth century we find that the chief influence on martial thinking and certainly the application of Sun Tzu to tactics and strategy was the cult of Marishi-ten whose inspired instructions were transmitted to the kenshi by means of her messengers, the tengu. This is an area of which I take the view that serious research needs to be done before the oral traditions that are widespread within some ko-bujutsu are lost irrevocably. Already we can note a marked reluctance on the part of non-Japanese scholarship to acknowledge the significance of the Marishi-ten, tengu (often referred to as 'goblins', a curious echo of the Buddhist demonization of these interesting 'beings') and yamabushi influences within the bugei. This is possibly through the lack of exposure at an advanced level within these disciplines of many researchers, although I am not sure whether this is not a weakness within the Japanese *shin-budō* field, as well.

Inextricably associated with transmissions within the upper level *kuden* revealed to some kenshi by these tengu are teachings that illustrate in simple but eminently sound terms a number of Sun

Tzu's most important principles. Some twenty years ago a prominent general of the Chinese Peoples Liberation Army (PLA), Tao Hanzhang, quoted Mao Tse-tung's dictum that 'Losing the initiative (in warfare) means to be defeated, to be annihilated.' While this was Mao's interpretation of the importance attached by Sun Tzu to the principle of gaining the initiative in any confrontation it was also well-understood by the Muromachi kenshi. Pre-emptive measures were essential in warfare; if possible, fighting a quick battle to force a quick resolution. Throughout the Iai-jutsu are to be found the exposition of mobility and flexibility.

One of the key strategies in fighting is 'making a feint in the east, then attacking in the west' to be found in the secret heihō transmissions and still considered of importance today. This is explained by Marshal Liu Bocheng of the PLA as the application of shifting from the 'orthodox' to the 'unorthodox', diverting the enemy as you wish, offering him bait which he must take and then striking him where he least expects an attack. Another example of developed understanding of the ancient Chinese master, clearly applied within the Japanese bugei as well as in modern times, was the triple principle of 'appraisal – the proper use of overwhelming force – and speed'. These two strategies, whilst appearing different, are really aspects of the same principle.

Without going into too much detail in this preliminary discussion, one of the key transmissions believed to have been revealed by Marishi-ten and reflecting much of the rationale above, is the application of the principle of 'circling' or 'spiralling' movements. By launching even the most basic of ch'i attacks using these 'coiling' movements, often with a barely-perceived shift to the 'east', the kenshi cannot be properly shaped, thus gaining a very slight advantage that enables him to strike in the 'west'. This inability to 'shape' the attack by the opponent can be said to be because the kenshi is 'invisible' – not physically or materially invisible but for the briefest moment of time, metaphorically effectively so.

Perhaps, when we probe beneath the surface and look for traces of the original inspiration that produced many of these innovative and clearly intuitive forms, we are beguiled to seek the rationale in the esoteric, particularly in Japan's case, the influence exerted by the Shingon mikkyō and, of course, the riddles posed by Zen. When we are pointed in the direction of Marishi-ten and Shugendō, we again tend to view these matters in terms of religious ritual theology. Sun Tzu expressed a clear warning not to allow military judgement to be clouded by either the deities or spirits, but this was something that, on the surface, the Japanese bushi appear to have ignored. We should take this question and consider, seriously, just how these kenshi gained their hard-won experience. Were they as religious as it seems, seeking their final inspiration from the deities? What was the nature of their maturing training that led towards their polished understanding?

One of the foremost characteristics of traditional bujutsu and budō is the necessity to undergo often ferocious training regimens. There is no easy way to gain experience and understanding, only grinding sustained practice throughout all the years of youthful vigour and eventual maturity. It is a common saying in Kendō, for example, that one's best years are in your fifties! Hardly the full flush of youth and implying at the very least twenty-five or more years of regular training sustained at the very least three times a week. Turn the clock back five-hundred years and that same period of hard training will mean hands-on practice every single day and prolonged for possibly five or six hours; practice continued from about the age of thirteen or fourteen for a young samurai of rank for a *minimum* of twenty-five years. Such skills contained in the major bujutsu required honing to a remarkably high standard.

The records tell us that young men like Aisu Ikkyō, Tsukahara Bokuden, Kamiidzumi Nobutsuna, and many others, at least once in their early careers embarked on a period of 'wandering the yamabushi paths' and so we should ask the question, what exactly did this mean? After all, these periods in the wilderness were not some sort of 'sabbatical year off' from their clan commitments, but a span that might very well last as much as ten years or more. Whilst one suspects that some of these budding kenshi did associate themselves closely with Shugendō and may have reached initiated *sendatsu* rank, there are no specific records that state this regarding any of them that I am familiar with. Nonetheless, they followed these 'yamabushi paths' and it increasingly suggests to me that during this time of what must have been considerable self-imposed hardship in the pursuit of their own private 'hell' of severe training, they associated themselves with those yamabushi groups who still retained the ancient skills with weapons that had been secretly transmitted from the early Heian era and possibly earlier. If they had wanted to gain battle experience and hone their developing skills by seeking out suitable adversaries with a redoubtable reputation, rather on the lines of Miyamoto Musashi in the early seventeenth century, there would have been nothing to stop them and this would doubtless have been recorded, but often such information is only coincidentally provided, almost as an afterthought.

It was during this long period of *musha-shugyō* when they 'polished their spirit' that their understanding of the principles of warfare took root. At length, when they felt that intuitive understanding was nearly within their reach, they sought another, briefer, period of seclusion that was very different from the first, and retired to the remoter parts of some shrine or temple in order to attain final perfection and to pray for the patronage of the shrine deity. These are some of the questions that remain to be answered but suffice it to say that these same kenshi of the fifteenth and sixteenth centuries, and some earlier masters whose names have come down to us but without any biographical details, focused their minds and their efforts on the practical mastery of the bugei, the arts of warfare, guided in the main by the genius of Sun Tzu. They sought their 'enlightenment' the hard way – not by sitting on their posteriors contemplating their navels. *Satori* may have come in their later years, just as it may have done so for priests and monks, but one feels that these warriors' lives may have been far more interesting. Part of their legacy remains with us today; we would be the poorer for its loss.

■ CHAPTER 1

The Muromachi Warriors and their Approach to Sun Tzu: Some Historical Benchmarks

Sun Tzu, in his work *Ping Fa*, appears on the surface to express his philosophy reasonably clearly and in such a way that in translation it is accessible to Western readers. This is somewhat in contrast to the other texts that make up the *Seven Military Classics* of China; however, except possibly in his 'Thirteenth Book' where the master discusses in the clearest possible way the employment of agents and spies, he carefully used language that was intended to veil the true meanings. His maxims and aphorisms were directed at prompting the intuitive understanding of those who commanded armies, the generals who had already gained experience in the field of battle and were masters of what they perceived as the proper tactics and strategy. His *Ping Fa* was a fresh appreciation of the arts of war based on his own knowledge of what had gone before and uniquely expressed in his personal philosophy. While ostensibly presented as a memorial to his ruler, he warned against the ruler being misguided enough to interfere or obstruct his expert generals in the conduct of military affairs where warfare was vital to the well-being of the state. At the same time his veiled language was not intended to be placed in the common domain and for very good reason. His philosophy of the arts of war was every bit as secret as the inner teachings of many of the bugei traditions that emerged in medieval Japan.

This is the nub of the problem in that the interpretation of this particular military philosophy has mostly been made, in modern times, by worthy scholars who undoubtedly have a deep understanding of the Chinese language and historical context, but who have perhaps lacked sufficient depth of background experience of the arts of war on the ground. The contribution made by these linguists in their translations and scholarly commentaries is not to be dismissed as irrelevant; on the contrary, they are significantly important, but it is the critiques presented by those scholars with a military background that appear to hold the centre ground. The fact remains that a true and proper evaluation of Sun Tzu's work can only be backed by long and hard-won practical experience whether in the processes of training integral to the bugei, or experience in the field. A very clear example of this combination lies in the writings of Mao-tse Tung where, in his instructions for the guidance of his field commanders, he presented his interpretations of the teachings of Sun Tzu in the simplest terms that could be understood by his unlettered leaders. He clearly realized that the ancient master throughout his writings often expressed his philosophy in general terms. 'Subtle and insubstantial, the expert leaves no trace.' What appears to be clearly stated is never clear, there are always layers of underlying meaning that only those who have gained hard-won experience are able to understand and even then not regard their findings as definitive.

Amongst the problems facing the commentator is that of attaining the proper background of experience. The specific field of military studies aside, a second approach is that taken by the Japanese kenshi during the Muromachi period. At once there is a serious barrier to be surmounted in the present day, that of distinguishing between ko-bujutsu and the more recent budo and, in particular, the all-pervasive dominance of shin-budo. It is essential to differentiate between the nature of those traditional systems known to have been formulated and brought to near perfection before the

Sekigahara watershed in 1600 and to have descended down to the present day with their integrity intact, those traditions that have been heavily influenced by the general 'softening' of the Edo period, and the many later entities that have entered the budo field post-Meiji.[1] The routine obfuscation by the media and the modern misconceptions of these Japanese warrior arts and ways will present great difficulties to any researcher lacking in-depth knowledge. A further problem is that through the secretive nature of these systems, and the older they are the more secretive they become, even the proper depth of personal experience will always be subjective and coloured by the teachings of a particular transmission. It is well recognized that few experts can ever fully master more than two or possibly three such traditions over their lifetime. By that, mastery means to be in possession of and fully comprehend the whole of the inner teachings.[2]

The necessary knowledge may be acquired should a researcher be admitted to any of these old systems of ko-bujutsu but then another hurdle will manifest itself. To reach deep understanding of such warrior transmissions it is usual to swear a blood oath that any teachings and secrets – the inner philosophical understanding – will NOT be revealed to anyone outside the tradition without the express authority of a licensed master or the incumbent headmaster of the ryu. This solemn oath made during the *nyumon-shiki* induction ceremony is as morally binding as, for example, signing the British Official Secrets Act in the present day. In medieval times, breaking such an oath by a warrior might place his life in serious jeopardy; today, such a moral commitment can all too easily be set aside as few modern students will truly subscribe to what is, after all, an archaic concept, long outmoded outside the shores of Japan. It is for this reason alone that it has always been a long and frustrating process for an aspiring student to enter such ko-ryu systems where all the old principles are strictly adhered to. It is also one factor that explains why such ko-ryu entities often have what appears to be a precariously small membership. They have always been rather parochial in nature, exclusive to one restricted circle of bugeisha practising under a succession of bloodline headmasters or under the aegis of cadet branches also jealously preserving their integrity. The admission of a prospective novice will often only be made after a lengthy vetting process designed to gauge the quality and character of the student and his family background.[3]

We can see this process clearly in the pre-modern induction of the *shingyaku* into Shugendo where the novice was required to undergo an arduous and prolonged period of severe austerities through 'entry' into a sacred mountain region, an experience far removed from his normal experience.[4] It is through such enforced collective experience that the 'wheat is separated from the chaff'. The privations imposed in Shugendo were very similar in nature to those prevalent in the late-Muromachi bujutsu and designed to prepare the initiate for the harder training that would come in this progression up through the successive levels.

Sun Tzu warned against placing faith on help from the spirits and deities. One may speculate if he really meant this literally or whether he was simply cautioning finding the balance between material experience and seeking inspiration by metaphysical means. The ancient world commonly sought the aid of a whole pantheon of deities who were thought to govern every aspect of life and the hereafter. It is at once apparent that there was the closest link between the material and spiritual world amongst the medieval Japanese bushi. Yamamoto Kansuke, (1493–1561), the great Takeda clan strategist who will be discussed later, once wrote: 'Throughout the year there are superstitions. However, in strategy (heiho) let your enemy be the one to be deeply influenced by the superstition, not you.'[5] However, there can be not the slightest doubt that metaphysics deeply influenced the medieval masters of the bugei.

Exactly what was the nature of these influences is a matter of some importance in understanding how the Japanese bugeisha interpreted the *Seven Military Classics*, especially Sun Tzu. The earliest may have been Taoist by nature through the undoubted contacts made by warrior migrants gradually moving east across the Sungari River into Manchuria and then southeast down into the Korean peninsular. These warlike Puyŏ migrant tribes probably came into contact with northern Chinese warrior-religious sects that seem to have existed in some regions of the northern territories on the fringes of the Chinese empire as they edged into Korea. By the second and third centuries CE these same people, now to be identified as Puyŏ-Kayan, began to cross the Tsushima Straits and forcibly settle in parts of western Japan. This was the start of the land-taking period that by the fourth century onwards to the sixth was characterized by the construction of many burial mounds known as *kofun*. These tribes not only brought their superior iron technology with them that soon established their dominance over the bronze culture they found in the archipelago but also their Taoist and Inner Asian shamanic beliefs that quickly assimilated many of the native inhabitants' animistic ones to eventually become what is now described as *shinkaku-shinko*. The animistic folk beliefs in what was mainly at that period the worship of a vast number of aspects of nature – storms, mountains, waterfalls, trees, strangely formed rocks, caverns, earthquakes and so on – might be termed 'primitive Shinto'. It can also be described as being the worship of 'Earth deities' and also included ancestral spirits, especially those of deceased ruling chieftains.

The newcomers who annexed the best farming land and soon subjugated, even enslaved, their predecessors, the Yayoi people, appear to have been led by sun-worshipping high chieftains who gradually justified their power by establishing a pantheon of 'Heavenly deities'. Memories of some kind of negotiated agreement forged in the Izumo region of western Honshu and recorded in the first two chronicles written in the early eighth century meant that the primitive 'Earth-deity' sankaku-shinko beliefs largely remained in what is now the western part of Japan with a clear demarcation approximately running from Ise in the southeast of the Kumano massif, then northwest through Kyoto to the Sea of Japan. To the east are to be found a predominance of the 'Heavenly-deity' beliefs. Of course, there has been a considerable overlap since those remote times and an adoption by either side of many important deities, but in general the above division seems to hold true. The beliefs firmly within the aegis of the shinkaku-shinko would seem, in later analysis, to have exerted the greatest influence on the developing warrior groups that began to come to a degree of prominence between the sixth and eighth centuries. In contrast, the 'Heavenly-deity' beliefs were more closely associated with the Yamato tribal groups and with justifying the group of powerful chieftains who soon became allied with or actually formed the imperial clan, The upsurge in building permanent shrines, probably at the beginning of the eighth century, on many sites that once were the main habitations of these Puyŏ-Kayan-Yamato magnates focused divine support from the deities to the 'right-to-rule' claims of the Yamato high chieftains.

Very early in this distant era, and certainly around the time that Buddhism was officially accepted in the mid-seventh century, Taoist mysticism was well established in a number of mountainous areas across Japan. The many ascetic mystics who withdrew to these sacred places clearly required 'protection' from the remnants of the indigenous tribes and from those who had fled into these fastnesses, probably eking-out their existence as marauding bandits, and this 'protection' was provided by the *proto*-yamabushi. Just who these 'yamabushi' were is open to conjecture but it is clear that they possessed martial skills that, a century or two later, became fused with a sort of religious asceticism that embraced not only sankaku-shinko but Taoism and some early form of the Buddhist mikkyo, the latter a century or more before its 'official' introduction at the beginning of the ninth century. The Taoist 'sorcerers' sought supernatural strength from the spirit world through their incantations but soon, in the popular mind, their yamabushi 'protectors' acquired the same reputed powers, possibly by association and by a very real ignorance and probable fear of the true nature of these occult rituals. Whatever was the nature of these supposed powers, it seems that these mountain 'priests' had already developed considerable skills with hand weapons, particularly the *naginata*,[6] the *bō* or staff, and the slung sword. In parallel with the development of the mystic sects came the absorbtion of two rather odd folk beliefs, the former to be observed in the cult of the *yatagarasu* and the latter, in that of the *tengu*. Both of these, and particularly the second, are significant in the understanding of the Japanese interpretation of the Chinese military classics.

The yatagarasu is represented as a three-legged crow and is worshipped either directly or indirectly at many shrines connected with those centred on the Kumano-sanzan. These shrines were also centres of Shugendo activity.[7] Shinto shrines adhering to the Kumano-sanzan, the three great taisha at Hongu, Nachi and Shingu, are to be found at many places throughout Japan but largely in southern Nara and across Wakayama prefectures. They are also located in the Izumo region of Shimane and Tottori prefectures. Whilst these sites enshrine a wide range of deities, these are mainly those deriving from the 'earth' kami despite a token presence of the sun deity or her relations, the 'heavenly' kami. One of the myths recorded in the earliest chronicles of Japan, written down early in the eighth century, was that Jimmu, the proto-emperor, was guided by a yatagarasu sent to him by Amaterasu, the sun goddess, when he landed with his men in the 'land of Kumanu'. Another legend, at the Kobe Ikuta shrine, recounts how Jingō-kōgō, returning to Osaka from her expedition to Korea, was guided by a yatagarasu and in a dream vision was told to land at Kobe instead, there to found a holy site that later became the Ikuta-jinja.[8]

Role of the Deities

Returning to the semi-mythical beliefs, the early warrior connections of the Kumano-sanzan are very clear. The commander of the first land-taking contingent who landed in Izumo was, according to legend, Take-mika-dzouchi, enshrined principally at the great Kashima-jingu in eastern Japan. It was this warrior chieftain who negotiated a 'peaceful' solution with the lord of Izumo, O-kuni-nushi,[9] and went on to 'pacify' the recalcitrant tribal leaders in central Honshu before finally seizing mastery and settling at Kashima and Katori.[10] Whilst the mythical and legendary aspects of the land-taking are set out in the two earliest chronicles, the *Kojiki* and the *Nihongi*, the fact remains that Take-mika-dzouchi and his brother deity or *alter ego*, Futsu-nushi-no-kami, preside over practice in all traditional bujutsu and budo dōjō, particularly those associated with the use of traditional weapons, right down to the present day.

In the Kumano region, always regarded as spiritual by the Japanese at large, Take-mika-dzouchi or his sword aided the Yamato clansmen in their difficult and hard-fought passage through the wild mountains before settling in the southern part of the Yamato plain, now southern Nara prefecture. It is in the legend describing the Kumano landings that we are told of the despatch of the yatagarasu by the sun deity, Amaterasu, to act as a guide. Legend also informs us that those who fell during this campaign were commemorated in two of the Kumano taisha that were later founded by the chieftain, Sujin. This chieftain is also credited with the foundation of the remote Udō-myōjin shrine that lies on the south-eastern coast of Hyuga province, Kyushu. To broaden the question of the yatagarasu further there is the possibility that before deification the yatagarasu was, in fact, a living person, possibly a minor chieftain. In the Shinto pantheon he is placed alongside Taketsu-numi and both regarded as sons of O-kuni-nushi, lord of Izumo. Be that as it may, the three-legged crow has long been used as a *mon* or crest, especially associated with the Kumano-sanzan.

Without going deeper into the whole question of the relationship between the deities now enshrined at the Kumano-sanzan and their 'messengers', suffice it to say that these shrines and the three-thousand or so associated jinja taking their mitama, have for more than thirteen hundred years been centres for constant Shinto and Shugendo mysticism and ritual. With the advent of Buddhism and the general Japanese leaning towards syncretism there has developed a blurring of the original connections; however, the yatagarasu also entered the inner beliefs of the bugei, probably through Shugendo beliefs, and their image is encountered in a number of *tora-no-maki* scrolls. This fusion of disparate beliefs, a conflated rationale, is to be found, for example, at the Kumano Hongu-taisha where the yatagarasu is described in present-day Japan, both as a messenger and as representing the three Kumano clans, the Ui, the Suzuki and the Enamoto. Whether these warrior families date from the Edo period or earlier is not clear. The symbolism of the 'three-legged crow' is stated thus:

Chi – Wisdom	Ten – Heaven
Jin – Benevolence	Ch'i – Earth
Yū – Valour	Jui – Humanity

These were said to be the virtues ascribed to Ketsu-miko-no-kami and Susanō-ō, the 'Storm deity'. From the violent nature of this deity he is often found housed a little apart from his fellow deities and regarded as a 'wild spirit' to be addressed circumspectly and with care. At some date the Kumano Hongu-shrine came to identify this rather terrifying god with Amida Buddha, a transition indeed!

The yamabushi element to be found at these shrines and their connecting roads is not so obvious but nonetheless a constant presence. Shugendo is a mystical religion that imposes severe privations on its followers. It is a mixture of esoteric elements from the Buddhist and Shinto mikkyo but also draws upon the primitive proto-Shinto beliefs that probably predate even the Yayoi period (300 BCE – 300 CE). Around the eighth or ninth centuries, the Nara and early-Heian periods, the cult of the female war deity, Mārīcī, was introduced from China although it did not originate there. This deity offered her adherants protection, especially in fighting, and it is therefore easy to understand her appeal. A number of her spells and rituals were soon absorbed into Shugendo. With the usual process of Japanization the name of this powerful deity was read and pronounced Marishi-ten in the islands of the archipelago. It is possible that the Mārīcī cult had found much earlier favour amongst certain warrior elements who accompanied the Yamato tribes and in practical form spread among the proto-yamabushi. Whatever the truth of this, it is an undoubted fact that Marishi-ten exerted an increasing influence on the emerging warrior class during the Heian period and was already fully recognizable in the Nara period yamabushi even before they were drawn under the umbrella of Shugendo.

The cult of this war deity, with its Shugendo overtones, becomes quickly apparent amongst the warrior groups by the close of the

Heian era, a period when pilgrimages to many religious centres, especially Kumano, became extremely fashionable with all levels of society from the bottom to the very top. Those who trod the several routes that composed the *Kumano-kodō*, the old Kumano roads, were described as ants, so many undertook the long journey throughout the whole medieval period. These pilgrimage routes were also termed the 'yamabushi ways'. One in particular was the Ōmine-michi that made its way south from the Kimpu-sen-ji temple at Yoshinoyama where is to be found the enormous wooden prayer hall called the Zao-dō.[11] From this Shingon-Shugendo temple the four seasonal 'entries' into the wild Ōmine mountains commence. The 170-km route follows a tortuous and difficult track southwards known as the *Ōmine-Okugake-michi* before ending at the Kumano Hongu-taisha. This route is also revered for representing the physical embodiment for the Taizoku mandala. For a yamabushi or a pilgrim to complete the whole circuit represents following both the Taizoku and the Kongo-kai mandala in the esoteric teachings of the Buddhist and the Shugendo mikkyo; in other words, to strive towards full perfection and enlightenment.[12]

The most powerful of the Buddhist 'Heavenly Guardians' is Fudō-myō-ō, the adamantine Defender of Buddha's Law, usually represented as a fearsome seated figure surrounded by cosmic flames and grasping in his right hand a naked sword to cut down those who would transgress the Law and in his left hand a rope to bind those who will submit to judgement. The severe ascetic practices that accompanied worship of these powerful images immediately appealed to the warrior who saw in them a sure route to gaining protection over adversaries through stiffening the mind and the spirit. The emerging warrior groups were fully aware of the martial skills contained within some sections of the Shugendo yamabushi and were fully embraced by the samurai, particularly by those who were forming the upper echelons, the buke. The mikkyo was seen as an immensely effective route to understanding the arts of war and in particular to obtaining enlightenment into the inner secrets contained in Sun Tzu and other military philosophies. The Marishi-ten cult within Shugendo held special appeal to the warrior and was readily absorbed into the very roots and foundations of the developing bugei.

When we examine in detail the Kumano cult of the yatagarasu and draw parallels with the distant Scandinavian beliefs connected with Odin as the Supreme Deity who emerged towards the close of the Migration period, we find the symbolism attached to Odin's two ravens, Hugin and Munin, is for all intents and purposes the same as that of the three-legged crow. The former stands for 'mind' and the latter for 'thought'. These ravens daily reported to Odin all that happened in the world. It is surely not by accident that this ancient concept of providing intelligence should appear many thousands of miles apart across the northern landmass particularly given that the Mārīcī cult spread in one form or another from its possible origins in Persia during the first millennium BCE.[13] Going through a number of changes the Inner Asian 'messenger' figure passed from a griffin, to a raven, to a hawk, a crow and finally to a tengu when it reached Japan, but always having the same function.[14] Essentially, this was symbolic of intuitive creative thought transmitted from the deity in a vision or dream and remained a constant amongst the skilled bugeisha for a span of at least five- or six-hundred years in Japan, that is, from the late-Heian to the early-Edo period.

Bugei Origins

One of the most serious errors that seems to have insinuated itself into the proper evaluation of the bugei in the last forty years has been the inability to place the origins within their historical context. Most commentators on the so-called 'martial arts' are heavily influenced by their varying degrees of experience within the modern shin-budo[15] and demonstrate a failure to distinguish properly between bujutsu and budo. This is despite a number of excellent books being published that clearly define the difference. As with many things at the present time the myth is far more believable than the fact and few practitioners of their particular budo discipline have the interest or motivation to search for the truth beneath the surface. It is unfortunate, too, that modern studies often seem to draw back from almost any reference to folk or religious influences, with the exception of Zen-Buddhism, and that the former produce a reaction of disbelief and the accusation of mysticism. In the case of Zen-do there is a totally uncritical acceptance of its 'all-pervading' presence in 'swordsmanship' through the writings of just one often-quoted authority.[16] A few serious Western students who have experienced at first hand and in depth some of the oldest surviving traditions of bujutsu are dismissed as holding extreme views and have, to revive an old colonial attitude that was prevalent in Western Europe prior to the Second World War, 'gone native' and their work is, therefore, somehow suspect. It is a matter, as pointed out by John Gray in reviewing a new study of Descartes,[17] that 'many contemporary philosophers believe that pursuing the sources of ideas betrays a generic fallacy, which wrongly suggests that if a belief has a particular origin it cannot be justified in other terms'.

One cannot visualize the Muromachi kenshi who interpreted the already ancient writings of Sun Tzu as men of letters. From a modern viewpoint they were relatively ill-educated even if we take into account their ability to compose quite sophisticated poetry. In their study of the Chinese masters of the Arts of War they drew their inspiration from their environment and sought the rationale based on personal experience. The fact that many of the great bugeisha who founded their own traditions followed these same 'yamabushi paths' must suggest something to the modern researchers. When we also understand that these same yamabushi paths and Shugendo, itself, were commonly known as the *'tengu-no-michi'*, the 'tengu-road', the truth begins to loom larger, does it not? It is irrational to make judgements on matters long past without trying to understand the environment in which these men lived. In order to reach back and try to grasp something of their thinking through the medium of their recorded actions it is necessary to totally disregard everything that post-dates their particular period.[18] History lies in the records of things past, the future is irrelevant in these matters. Only when we come to make critical value judgements on the impact of these martial philosophies both in theory and in practice does the 'future' enter the equation.

The great kenshi of the Muromachi era first studied hard and then tried to rationalize in practical terms the 'Thirteen books' of the Sonshi set against their personal experience of their troubled times. It was a period when they had every incentive to do so. Some attained deep insight but were content to rest their reputation on the strength of their newly formulated martial system; others found employment as masters of military tactics, recruited through the needs of every type of clan leader, large or small; still others achieved positions of considerable power as general commanders in the larger clan armies. These latter were the masters of strategy and their future lives depended on the effectiveness of their interpretations of the martial classics.

The great strategist, amongst several able generals serving Takeda Shingen, was Yamamoto Kansuke, already referred to above. We are fortunate to have what is generally recognized as a faithful redaction of his treatise, the *Heihō Hidenshō*, the Secret Book of Strategy. He probably compiled his notes just before or around the time he entered the service of Shingen-dono, in either 1545 or 1550 and completed the book by September 1561 when he met his death in one of the indecisive battles at Kawanakajima in Shinano province.[19] This strategist provides us, through his text, with a highly important insight into the thinking of some late-Muromachi kenshi. We are not reliant here on what may be the narrow transmission of techniques that may have been the subject of evolutionary change as they passed down the line of successive masters but have the actual thinking expressed in considered words by a kenshi of widely acknowledged repute, a master of the famous Kyōryū-no-heihō. Kansuke's succinct definitions of heiho are penetrating and worthy of close consideration.

The present study, focusing on the bugeisha interpretation of Sun Tzu (Sonshi) in Japan during the highly disturbed period that covered the breakdown of the Ashikaga Shogunate and the Sengoku-jidai, is an attempt to examine the roots that produced the martial brilliance that we can see in many of those ko-ryu that have survived to the present day. The key to this intuitive understanding rests with such admonitions as the words with which Yamamoto Kansuke ended his text: 'Strategy is something to be done with imagination.' These are the very words drummed into a few fortunate students even now, in my own case during severe training more than thirty years ago both in Kashima and Kyushu. In essence, bugei resides in the mind; first it is necessary to study and thoroughly master technique, then to allow that knowledge to mature before insight, or intuitive understanding, might possibly appear. Kansuke gave a further proviso: the student of these arts must add his own interpretations in order to advance; the result will be that each strategist will reach differing conclusions. The interpretations presented here are based on the principles contained in a number of ko-ryu whose roots were in the Muromachi period. In nearly every case in the Iai-jutsu traditions we find reflected the words of Ts'ao Ts'ao: 'Go into emptiness, strike voids, bypass what he defends, hit him where he does not expect you.'[20]

Focus on Practical Experience

Whatever was the nature of the many influences on the Muromachi warriors who sought to rationalize Sun Tzu's maxims these kenshi had one thing in common with the modern military commentators and that was firm practical experience of the theory as well as actual combat. Yamamoto Kansuke advocated that in order to become a strategist it was necessary to first acquire a basis in the unarmed martial arts, how to move properly and then to achieve harmony between the mind and body. Once these skills had been absorbed and, importantly, assimilated to such a degree so as to become second nature, then the student could advance to more specific levels, including an understanding of leadership, the use of terrain and the weather, to cite a couple of examples. This advice is not stated here as an isolated method, it sets out the structure found in all ko-ryu and onwards in the classical systems to the present day. Real understanding can only be acquired after many years slowly progressing up the levels, it never comes easily. Kansuke, himself, tells us that he had spent most of his childhood and all his youthful adult years studying heiho but even then he did not consider that he had mastered it. It was necessary to continue practical study all one's life. From this statement, evidently written when he was over thirty-years old, he had continued compiling his notes and observations for a further twenty or more years before drawing them together in his treatise on the Art of War.[21] Not only was it vital to attain insight in depth but also to become familiar with as wide a range of bujutsu skills as possible. Narrow experience in just one aspect was not enough. For this reason many of the greatest systems of heiho included a wide range of weapons, particularly as the student advanced beyond the basic *omote-do* level. Within the framework of military training within a clan, young samurai of rank would be made familiar with all four 'Pillars of Bujutsu': kenjutsu, sō-jutsu and naginata-jutsu, kyu-jutsu, and ba-jutsu, together with grappling in armour and a number of other areas, all in combination. Kansuke's teaching reflected the system prevalent in his day.

While another great strategist, Tsukahara Bokuden (1491–1572), is commonly credited with being the first master to recognize Sun Tzu's dictum that it was more important to take the enemy state intact rather than destroy it, and apply this to his bugei in general, surely Sun Tzu's evaluation of the objectives of warfare were well-known early in the sengoku period?[22] Perhaps the distinctively brutal character of medieval Japanese warfare in the sixteenth century was a barrier to understanding? The initial use of extreme and overwhelming force often resulted in difficulty in obtaining a peaceful and satisfactory conclusion to hostilities, a lesson not fully appreciated in our own time let alone Japan five or six centuries ago. It is of interest that according to the oral tradition within the Kashima Shintō-ryū, Tsukahara Bokuden may well have spent quite a significant amount of time in the company of Yamamoto Kansuke, either before or during the final phase of the latter's life, when he was an important strategist with the Takeda clan.

It is clear that these Muromachi kenshi realized quite early on the truth contained in Sun Tzu's dictum that 'he who understands how to use both large and small forces will be victorious'. Underlying many of the vast corpus of techniques that survive to the present day we can see the principle of applying intelligence to small operations that will bestow a disproportionate advantage to the weaker side. In these 'mini-engagements', for such is the nature of these military forms or 'kata', the kenshi learns to 'avoid strength and strike weakness'. When such understanding is employed in the field the strategist can see his opponent as though he stood on a lighted stage, brightly illuminated from all sides, whilst he, himself, is hidden from sight, able to move secretly, able to act on information coming to him from every quarter; subtle and insubstantial as a shadow. The Muromachi kenshi was undoubtedly a master of guerrilla warfare and this was the very essence of Sun Tzu's masterly teaching and the basis of the bugei.

CHAPTER 2

Who Were the Bugeisha?

To study the art of war in medieval Japan it is important to realize, at least from the end of the Heian period in the late-twelfth century, the time of the Gempei War between the Minamoto and Taira clans, the commanding generals in every group were men deeply conversant with the writings of the classical Chinese military philosophers known collectively as the *Seven Military Classics*. These works, and especially the *Thirteen Books on the Art of War* by Sun Tzu – Sonshi to the Japanese – dominate Japanese military thinking onwards through the subsequent medieval period to the present day. The questions that arise are how were the general principles contained in these works developed and understood, later to be put into practice in the field by these generals?

First, who were these 'generals' and how were they categorized by the Japanese warrior group, themselves? We can hardly do better than to turn the writings of one of the most famous of the sixteenth-century commanders who served the Takeda clan to the time of his death at Kawanaka-jima in 1561. This was Yamamoto Kansuke Yorinori who was born in Fujinomiya, in the Tōkaidō province of Suruga[1] in 1493. This master strategist, foremost perhaps amongst the several famous generals who served Takeda Shingen, wrote an important book on heihō that was eventually published many years after his demise. In this book, Yamamoto Kansuke defines commanders in the following words:

> **Heihō-zukai**: A strategist who has been able to master all that he has been taught but lacks the skill to apply the same in the field. One who can teach his students but cannot command.
> **Heihō-jin**: A strategist who understands the principles of heihō but has not yet mastered the teachings.
> **Heihō-sha**: A strategist who has mastered all he has been taught and can add his own interpretations to war skills. An expert and successful commander.

These military experts came to recognition by establishing their reputations through the study of bujutsu and eventually by understanding the principles of heihō that the various entities that comprised the arts of the battlefield demonstrated. The Chinese works mentioned above and a number of other works, now largely lost, formed the basis of the rationale that these men sought to understand and master. To do this, at an early age, they placed themselves, or were placed, under the tutelage of a warrior who was generally regarded as an expert in the field, and eventually reached a point where they mastered his teaching and were able to 'hive off' and develop their own interpretations of these arts.

From the beginning of the Kamakura period in the thirteenth century and certainly from the start of the Muromachi era in the first half of the following century, we find a growing number of references to martial traditions that we term today as *ryu-ha*. These warriors, generally known as *bugeisha* or *kenshi*, the latter term meaning a 'swordsman' in the broadest sense, recognized that in order to reach any sort of understanding of securing victory on the battlefield it was necessary first to gain practical skill (insight) into the nature and use of weapons and after this point to seek intuitive understanding that would bring invincibility and, very possibly, employment. This was certainly not a area for the faint-hearted or the clever academic who had learnt the Chinese texts by heart but had little or no experience of their application and the direction of large armies in the field.

The Seven Military Classics

Parts of the *Seven Military Classics* are thought to reflect military thinking from as far back, certainly, as the Eastern Zhou (Spring and Autumn) period (771–475 BCE), some scholars believing that they may contain parts of even earlier texts surviving from the late-Shang period before 1027 BCE when the Western Zhou was established. However, the majority, whilst probably reflecting earlier works, date from that violent period known as the Warring States era (475–221 BCE) with additions from the Han (206 BCE – 220 CE), the Three Kingdoms (220–591) and the Sui and Tang Dynasties (591–906). Pre-eminent amongst the works produced during the Warring States was Sun Tzu's *Art of War*.

For several hundred years, since the Sung Dynasty, the *Seven Military Classics* have been arranged as follows:

Sun Tzu's The Art of War
Wu-tzu
The Methods of the Ssu-ma (Ssu-ma Fa)
Questions and Replies between T'ang T'ai-tsung and Li Wei-kung
Wei Liao-tzu
Three Strategies of Huang Shi-kung
T'ai-kung's Six Secret Teachings

Whilst this is the generally accepted order, as pointed out above it may not represent the correct chronology. *The Art of War* and Tai-kung's *Six Secret Teachings* may contain much earlier principles, possibly originating several hundred years before the Warring States period. Some parts of the *Ssu-ma Fa* may date back to the early Zhao.[2]

It is clearly evident that Sun Tzu, once introduced to the warrior groups in Japan, the first texts possibly reaching the archipelago in the sixth or seventh centuries, had a marked influence on martial philosophy and the subsequent development of the whole of the bugei. Some authorities in the kendō field have long pointed to the earliest record of regular martial training being established towards the end of the Nara period in the eighth century, but other records hint at systematic swordsmanship being extant in the previous century, at least. Whilst this is not the place for discussion of the relative merits of such historical or proto-historical data, the characteristics of Japanese warfare from the late-Heian era in the eleventh and twelfth centuries are such as to suggest a considerable familiarity with the texts of the Chinese classics.

In these classical texts, and particularly in Sun Tzu, much trouble is taken to clarify the application of two aspects of the use of force. These are the opposing but interchangeable principles of the *orthodox* and the *unorthodox*. The former, in Chinese, is *cheng* and the latter is *ch'i*. *Cheng* may be rendered in English as 'orthodox' or 'regular' force and *ch'i* as 'unorthodox', 'unique', 'rare' or 'wonderful'. In ancient times *cheng* was used to 'hold' or 'fix' the enemy whilst *ch'i* groups were deployed to attack the deep flanks or the rear. Distraction was held to be of great importance as well as factors of time and space. When we turn to the Japanese bugei we might expect to find all these characteristics.

An example of Chinese practice being transferred directly to medieval Japan is the manipulation of flags and banners to obscure an army's 'order of battle' from the opposing side. In both ancient China in the Warring States period and in medieval Japan, armies were drawn up in mass formations around the commanding general. Each 'assistant' commander had his own banners and each formation their distinctive flags, well-known and easily recognizable to the opposing side. Groups of bannermen were constantly ordered to change position in order to distract the enemy commander and to prevent him 'shaping' the army. In other words, by the shifting of battle emblems in a constant to-and-fro repositioning, the enemy commander might become confused and the result was opportunities for *ch'i* operations.

Sun Tzu said: ' "Shape" the enemy but take every possible measure against being "shaped".' This is the principle of deception and underlies all military intelligence which is based on obtaining the fullest possible information about the enemy while at the same time denying such information to the opponent. Therefore, *cheng* is fixing and determining; *ch'i* is distraction which causes indecision and doubt. 'Who can tell where one ends and the other begins?'[3] The product of indecision, doubt and surprise is fear, and fear brings defeat. When we come to examining these principles as they are illustrated and given form within the Iai-jutsu, we shall find a number of excellent examples.

The contrasts between *cheng* and *ch'i* are directly correlative to the Taoist concepts of *yin* and *yang* – these should be considered in the order *yang* and *yin*. For reasons not yet fully understood, it is notable that Sun Tzu always put *ch'i* before *cheng* in his texts. *Yang* means 'the sunny (or light) side of the mountain' and *yin* the 'shadow' or 'shaded' side. The terms are inseparable, the use of the one requiring the mention of the other. In the bugei, for example, we have the sword posture *in-no-kamae*, 'shaded' posture that cannot be properly 'shaped' by the opponent, and *yo-no-kamae*, a 'clear' posture that (possibly) shows the intentions of the master. The latter forms are rarely used, the former constantly applied. This dualism conveys and contrasts 'clear and obscured', 'strong and weak', 'attacker and defender', and so on and so forth. It is to be found throughout the later text of Sun Pin's *The Art of Warfare*, 'lost' for two-thousand years until its recent discovery, but also in the teachings of his ancestor (grandfather?) Sun Tzu.[4]

This interchange between the orthodox or 'expected' and the unorthodox on the battlefield was brilliantly exemplified in early-modern European warfare by Marlborough's tactics at the battle of Ramilles against the French on 23 May 1706. The French, anticipating the conventional tactics that characterized most set-piece warfare of the seventeenth and eighteenth centuries, deduced from the Allied order of battle a flanking maoeuvre by Marlborough's cavalry, but the latter, whilst appearing to commit his horse in just such a move, launched his main but slightly delayed assault directly against the French centre. He thus outwitted the French commander and won a famous victory. This was the use of the orthodox as the unorthodox.

In order to use *ch'i* one must probe the enemy and find out his pattern of movement (Sun Tzu, VI, 21). To counter this one must be without ascertainable shape (VI, 24). This is the principle of 'invisibility'. By constant movement one gains the initiative, not by the use of aimless movement, but movement that cannot be shaped and will lead to the enemy being unable to properly defend everywhere. By maintaining the initiative one remains 'invisible' and able to strike because one is prepared. The enemy must maintain vigilance everywhere and endeavour to identify the place and time of the attack but is always 'one step' behind being able to do the same himself. This gave rise to the famous Japanese couplet often taught in the bugei:

> In quietness there is movement;
> in movement there is quietness.

Underdetermined Metaphors

In both the Chinese military writings and the later Japanese bugei the descriptions of military formations and, by extension, the names given to 'forms' within the ryu-ha are 'underdetermined images' intended to stimulate the student's memory so that he, and he alone, can fill in the details. The openness of the metaphor is often in total contrast to definition by language. This is one of the reasons why the forms listed by name in the bugei *tora-no-maki* or *denshō* can be translated largely as meaningful titles and yet only fully understood as the aspiring bugeisha progressively peels away the surface, layer by layer, to reach the real core. These are truly 'mouth to ear' secrets, only finally revealed through arduous prolonged training and final intuitive understanding. Each strategist or general will read into these principles his own interpretation; he exercises his own intuitive imagination in applying the principles to his specific needs.

Sun Pin further defined these opposing principles of *cheng* and *ch'i* by stating that the former can be used to dominate but should be coupled with the latter which has 'no form'. Like his ancestor, he states that the combination of 'surprise' with 'straightforward' operations produces inexhaustible possibilities.

T'ai-kung in his *Six Secret Teachings* advocated 'feigning and dissembling to deceive the enemy and allay suspicions'. It is interesting that his view was the waging of total warfare, a view endorsed in the very secret 'oaths' of loyalty required by many of the oldest Japanese *ryū-ha*. While some of the content may date from the beginning of the Western Zhou period (eleventh century BCE) the existing text comes from the Warring States; however, it probably reflects the oldest tradition of Chinese military thinking.

'One who excels in warfare does not await the deployment of forces . . . he manages them before they appear. He conquers the enemy by being victorious over the formless.' Anticipating the Japanese *'Four Poisons'* of bujutsu (and later budō), he states: 'Advance as if suddenly startled; employ your troops as if deranged . . . Who can defend against such an attack?'

The *Questions and Replies between T'ang T'ai-tsung and Li Wei-kung* quotes from Sun Tzu: 'Do not pursue feigned retreats' and 'Although capable display incapability'. Both are unorthodox strategies. Later, this work paraphrases Sun Tzu by stating that the combinations of the orthodox and the unorthodox cannot be exhausted: the one mutually produces the other in an endless cycle. What Sun Tzu refers to as 'giving shape to others but being formless ourselves' is the pinnacle in employing the unorthodox and the orthodox.[5]

The *ch'i* and *cheng* principles, directly reflecting the Taoist *yin* and *yang*, to put them in Sun Tzu's order, are to be found demonstrated in a number of the oldest surviving heihō-ryū[6] in Japan (late-fifteenth and early-sixteenth centuries). In the structure of the forms we find a large number of them complex and involved. There are many combinations of direct and indirect attacks coupled with defence that constantly flow between attack and defence and back again. These combinations within forms are often explained to the novice as the idea that he must understand that at any moment in warfare 'the fight may end in victory or defeat'. These forms, although carefully structured, are not stressed as hard and fast tactics but intended to broaden the swordsman's mind by fostering the concept of flexibility. Analysis, in the early stages of study, shows that in this flexibility there are both *cheng* and *ch'i* combinations. Whilst these forms are, on the surface, 'situational' they quickly bring the bugeisha to a high physical, mental and intellectual state of alertness thus preparing the ground for even more flexible understanding that will follow after a varying but always long period of thorough grounding.

■ CHAPTER 3

Ch'i Within the Eishin-ryū

It must be remembered that Iai-jutsu developed out of the perceived need felt by a number of mid- to late-Muromachi-period *kenshi* that the longsword could, in a number of situations, be drawn combatively, thereby giving the swordsman a distinct advantage of striking his enemy with devastating effect even before the fight had begun. According to the late Ozawa Takashi *Hanshi*,[1] the earliest Iai or Battō, to use the older term, were at first relatively unsophisticated but the principles of heihō developed greatly as the Sengoku period gave them urgency. With the introduction of Iai into some of the important early-Sengoku ryū, this 'branch' of swordsmanship soon focused a number of kenshi's minds on the principles already incorporated into Kenjutsu, Naginata-jutsu and Sō-jutsu in the specific circumstances where these close-space techniques might best be employed.

For the purpose of this study, I am taking the Iai-jutsu of the Hasegawa Eishin-ryū as my model as it contains a reasonable and varied number of developed 'forms' encapsulating the principles of heihō, but I wish to emphasize that the same extrapolations could be made from many more ko-ryū dating from the late-Muromachi period. No claim or assertion is being made that this particular tradition is the oldest or the best; it is simply that these are the forms I am familiar with through the valued teaching of my masters.

Whilst the Hasegawa Eishin-ryū is thought to date from around the middle part of the second half of the sixteenth century, ca. 1560–80, many of the techniques probably find their origins or inspiration in the mid-Muromachi before the Sengoku period. Analysis (Japanese: *bunkai*) of the three 'armoured' forms in the third, or *iai-hiza*, level (Urokugaeshi, Yama-arashi, and Iwanami)[2] suggests that the opponent is wearing armour of the mid-Muromachi period rather than the closer fitting and modified armour of the late-Sengoku decades. The argument rests on the nature and precision of the cuts made at the side and front of the opponent's throat, each followed by the dragging or raking 'pull' downwards or across with the turned blade of the sword. It is also thought by some masters within the ryū that these three forms, and possibly the next two discussed below, may have come from an earlier separate kata series designed specifically for armoured combat but now submerged within the later forms that have descended to the present day.

The two techniques (Tonarigaeshi and Namigaeshi) each have 'alternative' forms and correspondingly varying cuts after the draw. The former is generally interpreted as a form where the swordsman is 'seated beside another person' whilst the second form describes the technique using the metaphor of a 'curling wave or breaker'. But both are more subtle than these names would imply. In all the variations or alternative forms (kai-waza[3]), the intentions of the swordsman are masked by his 'curling' movement in both rising to deliver his *nukisuke* and in the whirling of his sword in cutting down.

While these forms are ostensibly simply rising and turning through either 90° or 180° on the 'draw', it is their possible inspiration that is the most significant in understanding the inner meaning. At the root of much, if not all, the Hasegawa-ryū lies the concept of 'invisibility'. This is implicit in the large movement whirling of the sword back past the left ear before delivering the *coup-de-grâce* (*kirioroshi*). This large spiralling motion is made in order to cause the opponent to involuntarily glance or shift his attention to his right, the swordsman's left, and thus render him less able to deal with the second strike. When we seek the original inspiration for many forms in the bugei then we should look for these in nature. It is from natural phenomena closely observed that many concepts are drawn. We need to remember, also, that Iai-jutsu forms are all delivered, in contrast to swordsmanship and the use of the spear or halberd, from a close interval where the slightest warning might shift the advantage to the opponent.

In these two forms, the inspiration may well come from the observation of spiralling smoke rising from small heaps of dead leaves swept up by the gardeners in the grounds of shrines or temples, a common enough sight all over Japan. It is not possible to precisely observe the 'form' of this smoke, even on the calmest of days, only to form an impression of its movement. While one cannot state categorically that this was the inspiration for both forms, nonetheless the swordsman's circling rise and cutting render him 'formless' and therefore *momentarily* 'invisible'; this coupled with a sudden vertical 'dropping' motion on the following cut, further shifts the advantage away from his opponent. Both forms, therefore, are excellent examples of the use of *'ch'i'* or the 'unexpected'.

It is when we come to the Oku-iai forms that we might expect to find these secret principles more fully examined and we shall not be disappointed. Ever since the Chinese military philosopher, Sun Tzu, so clearly set out the fundamental principles of the Art of War some two-thousand-five-hundred years ago, there has been a recognized distinction between the *ch'i* and *cheng*, the use of 'unconventional' and 'conventional' tactics on the field of battle.[4] It was the use of these two aspects that were so important in warfare, spelling the difference between victory and defeat, between life and death. The Iai-jutsu teachings contained within the Hasegawa Eishin-ryū and its several closely-related traditions are clearly the result of long study and the distillation of these principles. Just as the *okuden*, or 'secret' levels, in many other traditions seek to teach these truths to their initiates, so do these twenty-three forms with the addition of a number of alternative forms and the possible abandonment or loss of an unknown number of theoretical principles over the centuries of the transmission.

The origins, full description, and 'outer' and 'inner' meanings of each of the forms, are never explained, at least in the early formative years of training, but are passed down by each successive master from his intuitive understanding of the oral tradition. These are, above all, 'mouth to ear' matters and, by their very nature as well as by definition, so 'secret' as to be regarded as transmitted to the expert master swordsman from the deities, themselves. This is fully understandable if we consider how best we can describe, in clearly-stated terms, the creative process where a rationale has somehow been developed from an innovative intuitive flash of 'inspiration'. The Japanese are no better at describing how this occurs than we are, ourselves, but there is one fundamental difference. These are not 'cleverly devised' techniques created by some well-trained academic mind after a prolonged period of theoretical study, worthy as many such products may be, but entities often based on inspired practical knowledge gained in very real life and death situations. In most cases, the classical Bujutsu ryū were developed over a long period of time where the first master, the creator, was in mature years and had gained his understanding through bitter experience in the crucible of bloody combat. Very few such ryū were announced by young men in the years of their virile youth. The only explanation offered by these *so-dai*, or 'founders', and reasonably totally believed by them and their successors, was that the forms had been revealed by 'divine' inspiration.[5] One of the most frequently found attributions was that the secret teaching originated or was bestowed by Marishi-ten, the female deity of war, and transmitted by way of a tengu.

The 'Layered' Structure

Be that as it may, these 'okuden' levels encapsulate a vast amount of the distilled principles of warfare and very many of the techniques they contain are uniquely brilliant, but they all have one major and significant characteristic in common – that is they can *only be fully understood and evaluated* by a true master, one who has devoted his life to their study in the same manner as those masters who have preceded him within the ryū. In this aspect, understanding derives from the 'kami', the 'deities', and more particularly the *'uji-gami'*.[6] The techniques, themselves, are plain to see; their understanding can only come from long and serious exposure to their practice, and the successive re-evaluation that derives from 'peeling away' each layer to reach the one beneath, just as one might peel the skins of an onion; at the end of the process, each and every master might reach a slightly different conclusion as to what actually constitutes the 'secret' contained therein. Who can tell, since these are always oral traditions? Therein lies the vibrant life of many of these very old traditions, still creative even after the descent from master to master through twenty or thirty generations or more.

Hayashizaki Jinsuke Shigenobu (1546–1621(?)), born in the northwestern province of Dewa (possibly Sagami province), is generally acknowledged to have been the founder of the Hayashizaki-ryū and the inspiration for its several offshoots. It is thought, according to a tradition quoted by the Kendō historian, Yamada Jirokichi,[7] that this *kenshi* studied for a number of years and finally, as was the prevailing custom amongst such masters, secluded himself in a shrine for a long period in order to 'pray for final enlightenment'. He experienced no such phenomenon and decided to return to his home, presumably to renew his studies. Walking along the road in the dark, he had the feeling that he was being closely followed. The threatening feeling was so strong that he suddenly drew out his katana using his left hand, and stabbed backwards with a strong horizontal thrust – and killed a man behind him who was about to strike him down. This was indeed his *reimu*, or moment of 'divine understanding'.[8] The technique he is reputed to have employed is not now contained in the Hayashizaki Shigenobu-ryū or the Hasegawa Eishin-ryū but was passed to me through the teaching of my first Iai master, Takami Taizō, more than forty years ago and confirmed by the late Kamō Jisaku, ten years later.

It is in exactly this manner that intuitive understanding of the 'inner' theory is revealed. It is not a question of being instructed in a technique. Anyone can learn and perfect the physical movements involved, anyone can cleverly repeat a difficult name given to the technique itself and deliver what may seem to be a rational explanation, albeit one often couched in difficult pedantic language, but few practise long enough for the process of true assimilation to become part of the psyche *and be understood*. In other words, there are in Budō many expert technicians, that no one will deny, but there are very few who can be acknowledged as masters.

However superior we may feel in our view of the historical context, levelling criticism at the benighted medieval beliefs of the past, the fact remains that during the Muromachi era every single original style of the bugei[9] was thought to have been first revealed by Marishi-ten and transmitted to the founder by a tengu. In our enlightened twenty-first century we seek to explain the intuitive creative process by the use of modern jargon, sometimes intelligible but more often than not totally the opposite; in the Japanese bugei the attribution to some sort of divine revelation served exactly the same purpose but it also emphasized the fact that these beliefs were both mysterious and secret, the truth only revealed to those who were considered worthy of possessing the knowledge. We should not forget the point that in the Muromachi period there was a very urgent need to become a master and that 'edge' is now, for all intents and purposes, completely lost.

Personally, I make no apology for subscribing to the medieval *tengu-sho* concept of transmission. It seems to me, in consideration of the background and the inspired nature of many forms, to be every bit as valid an explanation as any modern attempt at definition through the use of anthropological semantics or psychoanalysis. It is also a good deal simpler, especially as we are dealing with a creative entity produced by relatively unlettered warriors. When students have reached the stage in their development, where several of the earlier 'skins' have already been removed, to use the 'onion' metaphor again, then the explanation of some 'secret' connection to the *Marishi* ⇨ *tengu* transmission often strikes an imaginative chord in their minds. They may not yet fully understand but they are very often much more receptive to fresh viewpoints.

The left-handed backwards thrust that may have constituted the moment of 'reimu' for Hayashizaki Jinsuke, is not now contained in the forty-five Iai forms of the Hayashizaki Shigenobu-ryū. One cannot say for certain why this should be so but the Jiki-Shinkage-ryū master, Yamada Jirokichi, pointed out that the process of dynamic refinement within the ryū-ha often was not completed until the time of the *san-dai*, the third headmaster in the descent. In the case of the Hasegawa Eishin-ryū we have the unusual fact that it was in the time of the seventeenth-century master, Ōmori Rokuzaemon Masatora, and the ninth head of the Eishin-ryū, Hayashi Rokudayū Morimasa, that an additional *ōmote-dō* level of eleven forms was inserted as an 'introductory' technical kata about one-hundred years after the ryū was first announced. The soundness of this new level is such that in modern times – throughout the twentieth century and to the present day – it has often stood as a separate entity on its own merit.

The fact that the backwards thrust form, ascribed to Hayashizaki Jinsuke by modern masters, is still remembered but rarely passed to students, embeds this difficult (and potentially dangerous) form firmly within the okuden levels and on a par with the last 'walking' form of the upper level, the *Shihōgiri* or *Ryū-no-ha*. Nonetheless, it is still a highly significant tactic when considered from the point of view of the number of other forms within the three upper levels that contain single- and double-handed thrusts; nine of the thirty-four waza. The form is also nameless, so far as I am aware, but it might serve to call it *Ato-tsuki* ('Rearwards thrust').

This drawing-sword technique is based on surprise and incorporates the unexpected. It is, therefore, a *ch'i* tactic *par excellence*. It has probably given rise, within this related group of ryū-ha, to at least four more forms, one of which is not named or contained in any of the regular levels. The swordsman, in drawing using the *Ato-tsuki* form, grasps the tsuka on his left step, holding the hilt with an underhand grip just below the tsuba (where the right hand would be if the sword were drawn), and draws forwards with the blade edge upwards on the following right step. As the blade clears the mouth of the scabbard so the swordsman thrusts directly backwards whilst simultaneously rotating back his left hip so that the blade punches back into the enemy but passes very close to the swordsman's left hip. He ends up half-turned in *hanmi*, having rotated back to the left 90° or more. The thrust will pierce the enemy just below the pit of the stomach.

CH'I WITHIN THE EISHIN-RYŪ

The immediate problem comes after delivering such a sudden rearwards thrust 'wrong-handed' because the swordsman is now left grasping his drawn sword with his left hand and not his right one. Unless he has trained himself to be ambidextrous, and there were masters who could do this, he was faced, in a situation where he was opposed by more than one enemy, with having to change his hand grip substantially in order to continue. It seemed to the late master of the Eishin-ryū, Kamō Jisaku sensei, that the *tachi-ai*[10] form, *Ukechigai*, developed out of the *Ato-tsuki* on the basis of arguing that the circumstances were more likely to be such that the swordsman faced a man ahead of him as well as a man close to his rear. The rear enemy, his awareness of a sudden threat by the swordsman visually hampered by the swordsman's body, would react to the threat and delivery of a rear thrust far too late and so be destroyed. The swordsman, on the other hand, by suddenly striking his front opponent, gains sufficient time to turn and thrust to his rear then turn back and cut down the surprised front enemy. The later form, *Monnu*, develops this *ch'i* concept further. Both *Ukechigai* and *Monnu* are supplemented by alternative forms.

In the Eishin-ryū the second form of the upper level okuden is another *ch'i* tactic. This is called *Zuri-tachi*, ('Sliding' or 'Companion' sword), and incorporates the properties of creating confusion, surprise, and 'invisibility' whilst dealing with two enemies almost at the same time, only they are now positioned one on either side of the principal, left and right, rather than front and rear. The thinking behind – *Zuri tachi* presupposes that the man to the swordsman's left is much more aware of potential danger than if he were to the rear. He constitutes the chief threat when the swordsman elects to act and so must be eliminated first. The problem is a spatial one. The swordsman needs to create space to allow himself unhampered action and in doing so threatens the man to his right by threatening to draw on him and hides his actual draw with his upper body swaying to the right and then the left (as he thrusts) and back again as he despatches the first man with a downwards strike. The momentum gained by the swaying actions and the large turning preparatory motion past the left shoulder and ear prior to the double-handed strike, all serve to enhance the transitory 'invisibility' of the swordsman. These swaying movements are designed to prevent his being 'shaped' by his opponent, one of the most important of the principles enjoined by Sun Tzu and one that forms a significant basis for this tradition's Iai-jutsu.

The second of the 'invisible' or 'non-named' forms that clearly belongs to the Hayashizaki group is another *ch'i* tactic that examines both an unexpected drawing motion and a foot movement away from the usual direction. This form might be termed *Hidari-nukiuchi* for want of a better description. Once again, we find that after this draw-cut the left hand is in the lead as opposed to the more 'normal' right. The hand positions may be 'normalized' as the swordsman turns in *jōdan* to show *zanshin*.

Invisibility

Perhaps the term 'invisibility' expressed as 'becoming invisible' or 'being invisible' to the enemy is a difficult one for the twenty-first century mind to understand? Viewed from within the Eishin-ryū and those classical bugei with which I have a passing but imperfect familiarity, the direct connection is that 'invisibility', or *shinobi*, is synonymous with the unorthodox *ch'i* of Sun Tzu. There is no suggestion here of the use of magical, cabbalistic, or occult rituals, although a number of such esoteric rites were closely associated with the mikkyō, only that the Japanese medieval mind sought to find the rationale. The term *shinobi* can mean the 'art of invisibility', that is true, but it can also mean to 'steal in' or 'infiltrate' and in a secondary sense, imply a 'spy' or 'secret agent'. The same meaning might be applied to *ongyoho* and *ma-jutsu* but these probably hint at the use of trickery such as a conjurer's or magician's skills of sleight of hand.

On the other hand, the warrior undoubtedly attempted to harness the esoteric mikkyō in a number of ways,[11] frequently resorting to the Taoist-derived formulae and rituals of Shugendō and Shingon. *Shinobi* is the term used to name two forms in the uppermost level of the Hasegawa Eishin-ryū, the one a variation of the other, which are what one might term 'night-time' techniques, ones requiring the swordsman to silently deal with an opponent whose presence at first he only suspects. Interestingly, the form, *ch'i* by definition, also contains a diversionary technique. At the end of the lower level of the okuden is another 'stealing-in' form, *Torahashiru*.

These forms have not the slightest connection with the so-called skills of the legendary Ninja of popular fiction but are based on the hard necessity of the warrior's requirements.[12]

CHAPTER 4

Hasegawa Eishin-ryū Structure

1. Ōmori-ryū

The Omori-ryū forms were inserted as an 'introductory' level by the ninth headmaster in the descent, Hayashi Rokudayū, somewhere around the time of the fifth Tokugawa Shōgun, Tsunayoshi, who ruled between 1680 and his death in 1700. This master studied Iai-jutsu under a Shinkage-ryū swordsman, Ōmori Rokurōzaemon. At the time when the forms were incorporated into the Eishin-ryū they were called the Shoden Ōmori-ryū Seiza-waza. From the point of view of the Hasegawa tradition, this is what is termed an *ōmote-dō kata* series of forms designed to teach novice *deshi* the basic principles of physical movement, regulated thinking, proper breathing patterns and, in the longer term, something of the principles that they will encounter later in the three upper levels. It will, in other words, bring the student to a reasonable degree of proficiency. According to tradition within the ryū, it was precisely the lack of combative understanding encountered by Master Ōmori amongst many of his young students in the late-seventeenth century that prompted him to devise this series. Whether or not this is true, the late Kamō Jisaku, a highly respected *dai-sensei*[1] whose dōjō was in Saga-shi, Kyūshū, passed the story to me; his approach to the classical ryū of both bujutsu and budō was uncompromising and he had little truck with modern sports concepts within the latter entities.

The order of the forms comprising the Ōmori-ryū are:

i. **Shohattō** or **Shoshintō**
This form demonstrates the four integral actions necessary in Iai and contains all the basic principles.
ii. **Satō**
'Right' sword although the sword is drawn to the left.
iii. **Utō**
'Left' sword although the sword is drawn to the right.
iv. **Tōtō**
'Rear' sword.
v. **Inyō-Shintai**
'Shade and Light'.
vi. **Ryūtō**
'Flowing' sword. Three alternative forms exist.
vii. **Kaishaku-tō** (now often termed **Juntō**)
The 'Sword of the Second' or 'Assistant' in *seppuku*.
viii. **Gyakutō**
'Reverse Sword'. Two variations of this form exist.
ix. **Seichu-tō**
'Controlling' or 'Restraining' Sword.
x. **Korantō**
'Tiger-violence Sword'.
xi. **Nukiuchi**
'Sudden Attack' or 'Draw-cut'. At least three variations within the tradition.
xii. **Inyō-Shintai**
Alternative form to 'v' above.

This 'alternative' form of Inyō-Shintai is always practised at the end of the Ōmori-ryū series although the reason why this should be so has never been explained to me. In the succeeding levels of the Eishin-ryū the alternative forms always follow on the 'regular' form.

2. Musō-Jikiden Eishin-ryū Iai-hiza Forms

i. **Yokogumon**
'Flat Clouds'; figuratively, the type of smallish cumulus clouds with flattened bases that one sees on a calm summer's day.
ii. **Tora-issoku**
'Tiger Leap' or 'Pounce'.
iii. **Inazuma**
Lightning (strike).
iv. **Urokugaeshi** or **Ukigumon**
'Scale Stripping' or 'Floating Clouds'.
v. **Yama-arashi**
'Mountain Storm'.
vi. **Iwanami**
'Rock-amidst-breaking-seas'. An alternative form also exists.
vii. **Tonarigaeshi**
'Dealing with a neighbour'. At least two alternative forms practised.
viii. **Namigaeshi**
'Recoiling Wave'. At least two alternative forms practised.
ix. **Takiotoshi**
'Water-cascade'.
x. **Battō**
'Close-Drawing' Sword.

The structure of this Iai-hiza level suggests that since the tradition was first formulated a number of forms may have been included from a separate set, possibly with the loss of others. The argument centres on Urokugaeshi (or Ukigumon), Yama-arashi, and Iwanami.

3. Okuden – Iai-hiza Forms

i. **Kasumi-tō**
'Mist' Sword. An alternative form is not usually taught.
ii. **Hiru-gakoye**
'Turn-aside'.

iii. **Togume***
'At a Threshold'. More than one alternative form exists.
iv. **Towake***
'In a Doorway'. More than one alternative form exists.
(* Some masters put these two forms after **Shihoniru** so that they become iv. and v. respectively).
v. **Shihoniru**
'Four-directions'. An alternative form is practised and interchangeable.
vi. **Tanashita**
'Below-a-Lintel' (or 'Below-a-Verandah Roof).
vii. **Ryosume**
'Narrow-space'.
viii. **Torahashiru**
'Tiger-Stalking'. Two alternative forms exist.

4. Okuden – Tachi-waza Forms

i. **Ukizure**
'Floating' or 'Escorted' Sword.
ii. **Zuri-tachi**
'Gliding-Sword' or 'Companions' Sword.
iii. **Sōmakuri**
'Sweeping-aside' or 'All Around' Sword. An alternative form is also practised.
iv. **Sōdome**
'Quell a riot' or 'Stopping' Sword.
v. **Shinobu**
'To Steal-in'. An alternative form is also practised.
vi. **Ukichigai (I)**
'Separation' or 'Passing-by' Sword. ('Dealing with a situation from the 'inside'.)
vii. **Ukichigai (II)**
An alternative form to vi. but practised in immediate succession, not as a adjunct.

viii. **Sodesuri-kaeshi (I)**
'Brushing Aside with the Sleeves'.
ix. **Sodesuri-kaeshi (II)**
Alternative form to viii. but practised in immediate succession, not as an adjunct.
x. **Monnu** or **Moniri**
'In a Gateway'. Two forms exist of this tactic but only one is usually shown.
xi. **Kabezoe**
'Against a Wall'.
xii. **Ukenagashi**
'To Ward-off'.
xiii. **Shihogiri** or **Ryu-no-ha**
'Four Directions' or 'Summing-up'. This form is often only revealed at the highest level.

At this point in this upper level three more forms are inserted but giving the impression that they are 'tacked-on' simply for want of another place. All three are *seiza*, or kneeling, forms and focus specifically on the formal *rei*.

xiv. **Itomago'i (I)**
'Leave-taking'.
xv. **Itomago'i (II)**
xvi. **Itomago'i (III)**
These three forms can be practised both from seiza and iaihiza, or half-kneeling.

Alternative Forms

Earlier in this study the expression 'Underdetermined Metaphor' was used to show that the terminology applied to describing military tactics and, by extension, the forms found in the bugei, was a device whereby the actual meaning was 'hidden' from those who have no need to know. When we look in detail at the structure of the Ōmori and Hasegawa Eishin-ryū and total up the number of *known* alternative forms within the framework, we find that there are as many as twenty-six forms. This is the equivalent of about two complete groups, half of them being within the two okuden levels alone.

The argument that forms may have been added as in the case of the Ōmori-ryū or that a level was somehow lost over the course of three centuries, surely cannot hold water since this was an 'official' tradition within the Tosa-han and, therefore, as was known to be the case elsewhere, most strictly controlled (supervised) by the clan councillors. They would hardly have countenanced alterations and additions creeping in.[2] The Tosa-han, along with the Shimazu and the Aizu-han, was noted for maintaining its military preparedness throughout the long Tokugawa period.

These 'alternative' forms do not seem to exist in some sort of limbo, some having an active rôle in the dōjō training, but usually they are produced, if that is the correct term, to amplify teaching at an advanced stage in the student's career. They are by no means 'irregular' by nature and, therefore, somehow diminished in their martial value. It is also clear that most, if not all, have been developed early on in the life of the ryū. I gathered the impression when I practised some two decades ago under a senior master of the Eishin-ryū in Karatsu-shi, Saga-ken, that some of these kai-waza forms may have been regional in their origin. Whether I am correct or not in this assumption I cannot be certain; nonetheless, they do exist and their function is to underscore both the validity of the 'regular' technique and to amplify it often by suggesting a slightly different application. We shall examine some of these alternative forms in greater detail later.

One striking feature of all this group of forms is that they do not seem to have been given a name that distinguishes them from the 'parent' form. The only exceptions appear to be Tora-issoku and Hiru-gakoye, neither of which appear to be 'alternatives', which are the second form in successive levels, yet hardly differ from each other in the manner that they are practised. The late Kamō Jisaku made no distinction between these two forms but he insisted that Hiru-gakoye must be practised faster and sharper than Tora-issoku. It is a question that remains to be resolved.

■ CHAPTER 5

Mao Tse-tung and Unorthodox Tactics

The military writings of Chairman Mao have been largely ignored in the West except among some Sinologists and military historians; in the modern world of Budō, they are probably almost universally unknown. This is a pity since Mao was a man with the widest experience of difficult warfare and who was able to interpret the ancient principles of Sun Tzu and others in a manner that could be understood by great numbers of unlettered peasants. He is believed to have memorized the entire text of Sun Tsu's *Ping Fa* as well as studying in depth the nature and course of such conflicts as the T'ai P'ing rebellion in the mid-nineteenth century and the Boxer 'Uprising' at the end of that century. His importance rests on his analysis of mistakes made in the past, and the remedies and re-appraisals that he instituted whilst fighting the Chinese Nationalists of the Koumintang and the Japanese successively, from the end of the first quarter of the twentieth century to the conclusion of the Peasants' War or Revolution in 1948.

Mao gained his experience the hard way as a fighting leader, a general who, whilst sometimes suffering serious reverses, ultimately won the war. In 1937, he ink-brushed this dictum: 'The first law of war is to preserve ourselves and destroy the enemy.'

MAO TSE-TUNG AND UNORTHODOX TACTICS

He was a master of guerrilla action, small or large in scale and, if there is a cogent reason why he should be studied, just take an overview of the result of the Vietnam War and at the conduct of a number of conflicts fought since. All these episodes were largely influenced by Mao's treatise *On Guerrilla Warfare* (Ch. *Yu Chi Chan*) published in 1937.[1]

Quite apart from his political objectives and his recommended methods of stirring and organizing the peasant masses, he couched the time-honoured principles of warfare in simple, easily understood terms, returning time and time again to paraphrase Sun Tzu and to develop practical understanding in his largely uneducated troops. Not for them the high-flown phrases of philosophers, the complex technical jargon of military colleges, the convoluted discussion of the principles that the vast majority could never understand. Simply, he made use of such plain metaphors as 'Threaten in the east; strike in the west'. He was like Napoleon whose dictum was 'ride to the sound of the guns'.

The guerrilla groups constituted the 'unorthodox'. They formed and struck without warning, and at once melted away. From their very nature such groups were 'formless'. They were everywhere and nowhere; they had no supply lines, no bases, no establishment. They were also able to gather information, intelligence, that made them largely the 'eyes and ears' of the regular commanders. All this is set out in Mao's pamphlet and it is recommended as required reading for any serious student of Bujutsu and Budō. For the purposes of the present study, the military writings of Mao Tse-tung are fundamental to appreciating the significance of Iai-jutsu. In analysing almost any area of the Bugei and particularly tactics and strategy based on Sun Tzu as they are illustrated in Bujutsu, Mao's comments and advice are of major importance.

The secrecy that surrounds the medieval Japanese heihō is fully understandable and from some perspectives was highly successful since their inception in preserving the core principles from outsiders. Yet it was Mao who levelled a telling criticism at the Japanese leaders in the period of colonial expansion from 1937 to 1945, that despite their long study of the *Sonshi* they had failed to understand it. One of the few dissident voices was Admiral Yamamoto Isoroku who pointed to the flawed policy just before the attack on Pearl Harbor but was ignored, thus bringing the USA into the war and the ultimate military defeat of Japan. A prophet is rarely heard in his own land.

Heihō not only taught practical concepts of the 'unexpected' or 'ch'i' in action but also demonstrated the principles of 'probing' and 'shaping' the enemy, the bones of military intelligence and its corollary, counter-intelligence. In probing, heihō gave insight into the strengths of the enemy, his dispositions, his supply lines, his morale, the state of mind of his commanders, their resolve or lack of it, his discipline, and the quality of his training. Every aspect of the enemy's condition was known, and in the greatest detail possible. Sun Tzu's much deprecated 'Thirteenth Chapter' concerning the use of spies and agents is as valid today as it was two-and-a-half-thousand years ago. It must surely underlie and reinforce classical Bujutsu even if it is largely ignored in modern Budō. 'First the eyes, then the feet, lastly the cut' is a famous maxim in both Heihō and later Budō; here particularly Kendō. By the 'eyes' the strategist means the gaining of information, the appraisal of the enemy. Another well-known maxim in swordsmanship is 'if we see a chance it is too late (to act)'. We may 'shape' our opponent but without the edge afforded by earlier intelligence, the chance based on his 'order of battle', his disposition, may fade away. Mao Tse-tung realized this and its application can be seen in the many 'small' wars of the last fifty years where 'guerrilla' or 'irregular' fighting has been a major factor. This is the basis of Iai, as one example, but the question remains that while the principles may be understood, their application may be hampered by the lack of vision by the commander.

The first masters of the many ko-ryū diligently sought their inspiration, not for the techniques, as that was relatively simple for any imaginative creative mind, but the 'revelation' that would provide them with the key to understanding. To experience this catalyst and recognize it when it appeared, was of supreme importance and possibly provides the reason why it is so difficult for most to understand.

Sun Tzu described in detail the failings found in many generals, Mao attempted to educate his commanders from the lowest levels upwards to avoid these weaknesses; the Japanese *heihō-sha* did their best to do the same. When we read the statistics that a famous heihō-sha like Tsukahara Bokuden (1490–1582) fought, during his long lifetime, seventeen *shinken-shobu* (live-blade matches), his first at the age of seventeen, several hundred *bokken-shobu* (wooden-sword) matches, and participated in thirty-six fully armoured battles, yet was wounded only six times by arrows in battle, we cannot put this all down to a very long run of good fortune. If this had been the case few prominent warriors of high rank would have sought him out for the value of his instruction. These men were realists and came to him recognizing the superiority of his heihō. He was, for example, invited to lecture on tactics and strategy by no less than Takeda Shingen amongst many others. To see such inate skills merely on the surface is superficial; it is also, in warfare, fatal.

Iai-jutsu, when considered from the point of view of combat conducted at a very close interval, is radically different from a match starting from a greater distance and more conventional dispositions. To fight at all requires contact with the enemy but to engage within immediate reach of each other is much more difficult; the slightest warning may precipitate a pre-emptive response from the opponent. The Iai-ka, therefore, must close to reach an effective distance without alerting the opponent and then, himself, initiate or pre-empt any response. This is the reason why so much emphasis is placed on developing the student's awareness coupled with his breathing rhythms *before* drawing out the sword. This is the critical time for a final analysis of the opponent's outer 'body-posture', reading his 'inner-mind', his state of preparedness; in other words, the assessment of intelligence and the denial of similar information to the other side.

Distraction and Concentration

General Griffith quotes Mao remarking that guerrillas must be expert at running away. They avoid static dispositions, strike where and when the enemy least expects them. They must lay baited traps in order to confuse, move without 'shape', distract in one direction but deliver their blow in another. 'Running away', according to Mao, and endorsed by Griffith, 'is thus paradoxically, offensive'.[2]

It is an interesting point to consider how to deal with a direct attack of a cut to the head, the basic and often stressed most important attack in swordsmanship. In modern Kendō developing out of classical-Kendō, (i.e. post-1600), the 'defender' seeks to avoid the attack, often by stepping back out of the path of the attacking blade. Alternatively, he will seek to pre-empt by delivering his own attack. Disregarding the latter course which brings in another principle altogether, the fundamental idea is 'avoidance'. However, if we go back to the application of the older heihō to such an attack, the swordsman does not avoid backwards but often steps forward *under* the attack. It is safer to step closer to the attack than to avoid backwards. There are very good reasons for doing this. The first of these, on the physical level, is that the attack is probably focused on the precise judgement of distance, in this case the need to strike with that part of the blade known as the *o-mono-uchi*, the area between three and six inches, (eighty to one-hundred-and-fifty millimetres), from the tip. To avoid the trajectory of the cut may be fine but if it is not judged exactly there is still the greatest danger of injury or death. By stepping-in and dropping under the cut, the risk of serious injury is diminished as that lower part of the attacking blade is not intended to be an integral part of the attack; it is also moving slower from the point of view of the laws of physics. Because of these factors it has a reduced penetrating power. More importantly, coming under the opponent's attack opens up the chance for a range of *ch'i* counter-attacks which, by reason of being 'close-space', are devastatingly effective.

This practical example from old Japanese swordsmanship and, to some extent, the use of the short-shafted spear and halberd, reflects Sun Tzu's all-important principles of 'distraction on the one hand and concentration on the other; to fix the enemy's attention and to strike where and when he least anticipates the blow'.[3] We return here to *cheng* and *ch'i*, but the flexibility implicit in such tactics means that *cheng* is not wholly the orthodox and *ch'i* is not entirely the unorthodox. 'It is only the wise general . . . who is able to recognize this fact and turn it to good account.'

To fall back and avoid is to disperse. It is a tactic that can invite destruction. In classical Kendō, i.e. swordsmanship developed after the mid-seventeenth century, and the modern entity this form of avoidance followed by a counter-attack has been termed 'negative' Kendō; in classical Bujutsu to step in under the attack is thought of as 'positive'. The classically-trained swordsman when faced with such an avoidance would at once renew his attack without the slightest delay. This is the principle of taking the initiative and keeping it, not to allow the slightest chance of regrouping after avoiding the initial attack.

In the Hasegawa Eishin-ryū Iai we are considering, there are a number of forms in each of the four 'levels' that address and illustrate this principle. In 'positive-action' swordsmanship the use not only of horizontal movement but also vertical movement is applied to this same principle and may be equated with the Chinese maxim 'sheng-tung, chi hsi' – 'uproar (in the) east; strike (in the) west'.[4]

Mao patiently explains in simple terms his concept of winning people over to his cause, in this case mobilizing them to take on the rôle of guerrillas. His method is exactly the same as in the classical Bugei and, possibly, Budō. *'Explain'*, *'persuade'*, *'discuss'*, *'convince'* – these words occur again and again.[5] It is a restatement of the age-old principle of 'start with the simple; progress to the more difficult'. A master will only explain gradually the deeper meanings or, returning to our earlier metaphor, peel away the successive layers of an onion, once the student shows that he has grasped the basics. One of the most influential Iai teachers instilled a conviction that the Iai-jutsu technique 'would be effective' *because this was how he taught it*. The statement brooked no argument; it contained no hint of weakening in the belief simply on the grounds that the student had not tested it by *tameshigiri*.[6] He would say: 'Why test it? If I say it will work, it will certainly do so. But it will not work if you question the premise!' Mao's progression from 'explain' to 'convince' is right, of course; in modern parlance 'convince' is self-belief. He meant it in a broader context of collective self-belief, but to the warrior it is of utmost importance since action and survival depend on total commitment.

Near the beginning of Mao's *Yu Chi Chan*, which he wrote in 1937, he outlines and rephrases the basic principles of Sun Tzu with particular reference to unorthodox (*ch'i*) tactics:

> In guerrilla warfare, select the tactic of seeming to come from the east and attacking in the west; avoid the solid, attack the hollow; attack; withdraw; deliver a lightning blow, seek a lightning decision. When guerrillas engage a stronger enemy, they withdraw when he advances; harass him when he stops; strike him when he is weary; pursue him when he withdraws. In guerrilla strategy, the enemy's rear, flanks, and other vulnerable spots are his vital points, and there he must be harassed, attacked, dispersed, exhausted and annihilated.[7]

This passage could easily have been written with the substitution of 'Iai-jutsu' for 'guerrilla warfare'. It exactly parallels the principles of Iai within heihō. In the *Ping Fa*, Sun Tzu wrote: 'Appear at places to which he (the enemy) must hasten; move swiftly where he does not expect you'.[8]

■ CHAPTER 6

Iai-jutsu Seen as Flexible Warfare

As we have seen, Iai-jutsu was, within the heihō, 'flexible' warfare. By the term 'static' we understand regular or orthodox tactics and strategy, whereas flexible implies the use of unorthodox methods. In static warfare, the fighting units are generally deployed with a front and a rear, with similar structures on either flank in support. In flexible warfare there are generally no fronts or rears, and frequently no flank support. Iai reflects this and by its nature seeks quick decisions.[1]

When we observe many of the older entities of classical Bujutsu we often see examples of relatively long and complex kata structures where the principles of attack and defence ebb and flow through a number of conjoined 'mini' engagements ranging from two or three segments to a much greater number. In Iai-jutsu, while we occasionally have such distinct but linked moves, these always seek a quick rather than a protracted outcome. The swordsman may have a 'front' but he has no recognizable 'rear' or 'flank'. So far as he is concerned, he fights without support. These are exactly the principles of mobile warfare reduced to its lowest common denominator. This is in contrast to much of other Japanese swordsmanship where the practice of these principles is often presented as 'combat between two generals', oneself and the opposing commander. It is also why the development of *zanshin*, 'awareness' or 'remaining mind', is so stressed in both physical and mental alertness. This is termed *zanshin-no-jutsu* but as this state develops within the student, it deepens to an 'internalized' level known as *zanshin-no-ri*. At this more advanced stage of innate 'awareness', zanshin is fused into the student's very being, becomes one with the understanding and evaluation of intelligence and, naturally, the application of counter-intelligence. While this holds true in the more static situation, in a mobile *ch'i* confrontation accurate appraisal becomes essential.

In many Iai-jutsu forms, the opponent (or opponents), can be likened to a static or regular force whereas the swordsman is a mobile force. The student trains himself to recognize the circumstances that will allow him to move at the exact moment that he judges he has the initiative 'To know and to act are one and the same', is an oft-repeated maxim. It means 'to act on sure knowledge'; in Bujutsu terms, to practise repeatedly each and every form until absolutely perfect and the principle underlying the technique is thoroughly absorbed. Only then can the mind freely work so that action becomes intuitive.

Gaining the initiative is essential for the application of *ch'i* tactics and achieving victory, but once the action becomes protracted then the student is forced back into 'static' warfare. Without support and with the element of surprise lost, he will be vulnerable. Unless he can 'melt away' or withdraw in a manner that gives him strength, he risks defeat. It is for this reason that most Iai-jutsu actions are short, explosive and devastating; they always aim to reach a rapid conclusion.

It is the fluidity stressed by Sun Tzu and explained succinctly by Mao Tse-tung that is the essential element for success. 'One of the characteristics of mobile warfare is fluidity', wrote Mao.[2] Zanshin develops the ability of 'weighing the situation', being able to 'know the enemy', to read the enemy's mind, then to strike. Sun Tzu, crystal clear as ever, encapsulated this in the simplest language: 'Weigh the situation, then move.[3]'

In this study of heihō the aim is to illustrate and explain the principles contained in Sun Tzu and some of the other military classics in terms encapsulated in such entities as the Iai-jutsu forms.

Many of these particular forms demonstrate the late-Muromachi understanding of *ch'i* as opposed to *cheng* tactics, and the fluid combination of these in their myriad ways. The Japanese masters in this period were realists and fully realized that every aspect of military understanding was going to be tested in the crucible of the battlefield; therefore it was vital that they came to the correct conclusions before announcing any innovations to the warrior world at large. 'If a theory, expressed as a "form", did not work under such conditions, it was discarded immediately.' This caveat is frequently repeated in classical Bujutsu even today. The seriousness of such principles failing to work was minimized by the fact that practice was reduced to its lowest common denominator, as has been pointed out before. Conception, practice, testing, adjusting, allowing for the different factors of circumstance, and so on and so forth, all contributed either to the honing to perfection of the final movement or its eventual rejection. Often it must have been found that what worked for one master did not for another. To use a modern metaphor, 'the proof of the pudding is in the eating'.

The Chinese Arts of War as Interpreted by the Bushi

I think it is important to keep in mind that our interpretations of SunTzu's *Ping Fa* are chiefly based on the modern translations of the *Thirteen Books* that are accessible today, and as such they may be subjective. Matters may not have been the same in ancient Chinese times and the work may have been viewed differently as the political-military situation worsened in medieval Japan. Lau and Ames point out that by the middle of the Han dynasty the *Ping Fa* had evolved or been added to so that it had accumulated some eighty-two chapters. Sun Pin's work, known generally as *Sun Pin's Ping Fa*, had grown to eighty-nine. Both works were composed, in all probability, of both an 'outer' group of chapters covering historical events that were included with the intention of illustrating a narrative approach, as well as the more specific 'inner' exposition so well set out in Sun Tzu. We can see this format in other works included in the *Seven Military Classics*. It seems probable that by the time these Chinese classics had reached Japan and aroused the interest of some early warriors, the teachings they contained would have been transmitted in the form of lectures but the texts, themselves, would have been highly valued and jealously guarded.[4] The report of the youthful Minamoto-no-Yoshitsune secretly copying out by hand two Chinese military texts, the *San Lueh* and the *Liu-t'ao* whilst he was living in the care of the Abbot of the Kurama-dera north of Kyōto in the last quarter of the twelfth century, suggests that at that time in the late-Heian period such treatises were rare and the availability limited.[5]

The custom of presenting these treatises on the arts of warfare couched in the form of dialogues between the 'master' and his ruler is understandable since the teachings were not intended for dissemination to all and sundry and that includes, in particular, other generals and strategists. These were certainly in the category of 'mouth to ear' secrets. When we examine this in medieval Japan we find much the same for the jealously-guarded denshō of heihō. Possibly the surprise should be that these works were written down at all in such oral traditions.[6]

The preservation of these secret matters clearly became a very important consideration for the kenshi in the Muromachi, the precise period that saw the breakdown of the power of central *bakufu* controls and a corresponding upsurge in lawlessness and military violence in settling disputes. It is probable that some warriors in the upper levels of the bushi class appreciated a significant decline in the Ashikaga bakufu and tried to maintain secrecy within their own sphere concerning these peculiarly military matters by imposing strict moral 'rules' on their students that may be considered the forerunners of later military law. If this is the case then there are almost exact parallels in late medieval Europe.

The founders of most, if not all, the great Muromachi period bugei traditions sought to preserve their newly conceived and polished concepts from all but their inner circle of deshi and even these disciples were not made fully aware of the deepest 'secrets' until they had proved their worth over many years. Even here there was a reluctance to impart such military knowledge that borders on that sinister maxim that 'all warfare is based on deception'. Often, the final revealing of the teaching was only passed from the headmaster to his successor at the very end of the former's life and marked by the solemn handing over of the tradition's *tora-no-maki* scrolls. Accounts of such 'deathbed' transitions abound.

This is not to say that there was not a partial dissemination of the inner meanings; there was, of course, but it was a slow progression clearly based on constant assessment of the suitability and reliability of the preferred student as he passed into each successive stage of awareness. In some *ryū* progress was often marked by the presentation of *denshō* scrolls, simple statements at first, but increasing in length and detail as time passed. A student might, after many years, be in possession of several of these documents.

Near the outset of a deshi's admission to a tradition, and akin to the *shingyaku*, novice, being admitted to the lowest stage of initiation in some religious 'societies', Shugendō being a good example, he would be invited to swear an oath of faithfulness, affirmed and signed in his own blood, that nothing of what had been or was to be revealed to him would be divulged to anyone under any circumstance, without reference and direct permission from the headmaster. These strictures were the medieval equivalent of the 'Official Secrets Act' in the UK and, like their modern equivalent, were considered binding for life.

The oath given so solemnly at the *nyumon-shiki* ceremony contains a significant reiteration of the opening lines used by Sun Tzu and Sun Pin that 'war is a matter of vital importance to the State . . . It is mandatory that it is to be thoroughly studied' (*Sun Tzu*: I, i.). *Sun Pin* (2) emphasizes that failure to understand victory or defeat can be catastrophic; 'military situations must be examined with great care'. The formula used in the Japanese oath implies the great significance placed on heihō as vital to the State. These are clearly matters that are not in the public domain. In two short sentences the initiate is told that he must *never* use any of the techniques (principles) in almost any conceivable circumstances other than *in extremis*. 'You must go to war only when there is no other alternative', wrote Sun Pin. His ancestor stated that war was 'the province of life or death'. The masters of the bugei interpreted this in similar stark terms, that nothing of what the student learned within the tradition may be used except in the most dire emergency, then it must be employed to the full with nothing held back. If the student was killed then nothing of the secret 'inner teaching' was lost; if he secured the victory then, conversely, nothing was given away.

The medieval Japanese interpretation implies a full understanding of the importance of these military secrets but, in practice, reserves to the discretion of the tradition's headmaster or his *fully experienced* senior deshi any decision on further dissemination.[7] This meant that the explanation of the principles, the heihō, might not be handed on to anyone, even their lord. These masters were the custodians of the secret teachings; they might make use of them in warfare but they might not hand them over. It was clear that the great lords, their masters, tacitly recognized this.

To secure such secret knowledge into his own hands, the ruler must himself become expert. While many did study the heihō diligently, famous examples being the three greatest figures of the late-Sengoku period, Oda Nobunaga, Toyotomi Hideyoshi and Tokugawa Ieyasu, all sought out and retained the services of brilliant generals and placed considerable reliance on them – and, doubtless, sent agents to watch over them. Ieyasu, a master of warfare by any standards, spent much time in his youth practising Bujutsu and was an accomplished swordsman in his own right; yet he once said that he did not need to master the use of the sword, he could employ other men to do that. The historical record shows us that he was a great strategist and won a number of famous victories based on his command of the principles.

There is another excellent illustration of this. A famous kenshi who was also a successful general was Toda Seigen. He was the founder of one of the two Toda-ryū-no-heihō. Seigen was once ordered by his lord to teach him the secret form of the Toda-ryū called '*Mute-kachi*', or 'Handless-victory'. Unable for obvious reasons to refuse such a direct order, Toda Seigen reluctantly agreed to do so but first obtained assurances that as this was so secret a teaching, he and his lord must ensure that they were alone and not overlooked. He demurred for so long that his lord eventually said that he would check their privacy in person. As he turned to do so, Seigen seized hold of his master's hand, thus preventing him from any possibility of drawing his sword. 'This is my "Handless-victory" teaching, my lord!' His master understood that in such matters as strategy the ruler has no business to interfere. Sun Tzu wrote: 'The side on which the commander is able and the ruler does not interfere will take the victory.'[8]

Returning to the perspective of the swordmaster in assessing his student's worth, he might demand a full two to ten years before permitting the *nyumon* ceremony of initiation. This period allows the master time to examine his deshi's determination and steadiness of character. However, full admission may take very much longer; fifteen or twenty years being quite common. This may seem quite harsh but it has a certain dark humour to it. It is a commonly heard pleasantry in modern Kendō that masters first really recognize their student's faces at about twenty years; they actually remember the same student's names after thirty years! The serious side of this is the need to be absolutely certain of reliability; that the student does not have some hidden moral flaw that could eventually surface and bring the tradition (or clan) to ruin. The ruler must have complete trust in his commanders.

'All Warfare Is Based on Deception'

If the premise that war is of paramount importance to the State holds true then it is of primary importance to the State that the leading commanders be masters of their art. Recognition of such mastery lies with the ruler and can be no easy matter. Examining medieval Japanese military history from the point of view of the bugei affords one most interesting example of this.

Yamamoto Kansuke Haruyuki (1490–1561) seems to have developed no particular claim to fame in his early years although he made a reputation for himself in mastering the well-known Kyō-ryū-no-heihō. He was lame in one leg and had lost the sight of one eye,[9] yet despite a blunt refusal to employ him by Imagawa Yoshimoto

(1519–60) in whose domain he resided, he was accepted by Takeda Shingen (1521–73) purely on a recommendation by friends. On arrival at Kofu, Shingen's capital, the previously agreed stipend was doubled by his new lord. Takeda Shingen evidently followed the early Chinese precedent in first assessing Kansuke's character but it was later that he found his trust justified when Kansuke soon proved his worth as a brilliant strategist.[10]

The necessity for careful evaluation both of a general and a deshi of heihō is that each, at his own relative level, has access to secrets that are vital to the state or the ryū. The importance of heihō in the Muromachi period, as much as at any other time, was that the individual master, the true expert strategist, learnt his craft in the manner advocated by Sun Tzu and understood that 'the control of many is the same as to control the few'.[11]

The heihō-jin was a complex figure not without some ambiguity. Shingen and his advisers in the inner council of the Takeda clan must have been well-aware around the middle decades of the sixteenth century, that Kansuke was a masterless military expert, not exactly a rōnin as he possessed his own land from which he must have taken an income, but nonetheless not as a retainer of Shingen's father-in-law, Imagawa Yoshimoto. As a master of heihō, Kansuke would have easily been able to evaluate the military preparedness of any lord through whose territory he travelled. His judgement would have been dispassionate and expert. Doubtless, too, he amplified his knowledge through a network of his own informants, certainly throughout the Kantō, the Tokaidō provinces and the central Chugoku region. He would certainly memorize every scrap of information that he gleaned.

Shingen probably agreed to retain Kansuke around the year 1540, the time when he revolted against his father, Takeda Nobutora, and seized power in Kai province placing his father in the custody of Imagawa Yoshimoto.[12] By this time, Kansuke was about fifty years of age and a mature warrior. The employment of this kenshi as a strategist ensured that he would not find a place under the banner of a neighbouring lord, particularly a potential threat like Imagawa-dono should he become ambitious through his family connections; the

Hōjō-ke in Sagami province; or the Matsudaira in Mikawa. Although ostensibly vassals of the Imagawa, these feudatories must even at this date have realized the strategic value of their ancestral domains at the western end of the relatively narrow coastal corridor along which passed the vital Tōkaidō highway. As a 'masterless' heihō-sha, Kansuke would have represented a valuable 'catch' for almost any ambitious lord at that stage of the Sengoku period.

Probably soon after Kansuke was retained by the Takeda he arranged for another great heihō-sha to travel to Kai and expound on tactics and strategy before his lord. This was no less a man than Tsukahara Bokuden, master of a small castle fortress in the Kashima domain at the lower end of the Kashima Hantō on the Pacific coast. Bokuden was the same age as Kansuke and it is quite possible that they already knew each other well and had practised together during the very long periods that Bokuden 'followed the yamabushi paths'.[13] There seems little doubt that there was a close connection between the Kashima-ke and the Takeda-han at this time, although at what level precisely is not clear, however, the tradition persists within the circle of the Kashima Shintō-ryū swordsmen.

The friendship contact between two such expert kenshi would give access to a huge fount of military expertise based on the principles of the art of war contained in the *Sonshi*. Both men had honed their understanding of these principles for more than thirty years and could be numbered amongst the foremost proponents of heihō in their day. Yamamoto Kansuke went on 'to greatly increase his lord's territory' and even before 1544 was instrumental in the capture of as many as nine 'castles' for Shingen, whilst Bokuden remained unattached but gave regular instruction for many years to some of the most important figures of the period including two of the last Ashikaga Shōgun and the powerful warrior-aristocrat, Kitabatake Tomonori (1528–76).

In a period when moral rectitude appears to have been in short supply, if the accusations of duplicity, double-dealing, treachery and naked ambition levelled at quite a number of the Sengoku daimyō are to be believed, it is remarkable to find that many of the kenshi were regarded as 'reliable' and 'faithful' generals. Whether this was the case or not, or that fettering hostage safeguards were imposed, is a matter for further research and analysis; certainly some were regarded with suspicion by their lords in the closing two decades of the Sengoku era when the final paroxysms of the unification campaigns were coming to a head, but it is notable that many of the founders and early successors of these classical ryū did enjoy a reputation that singles them out as exceptional amongst their peers. There seems to be a direct connection between their reputation and the blood-oath each swore, or instituted for their deshi, to uphold the integrity of their martial system.

Sun Tzu delineates the three ways in which a ruler can bring misfortune down on his army. All three refer to interference with his commanders.[14] The history of the Takeda-han clearly shows that Shingen had a number of completely faithful generals and that he heeded the advice of these vassals including strategists headed by Kansuke and did not interfere in their plans.

The contrast came after Shingen's unfortunate death in 1573 when, two years later his third son, Katsuyori (1546–82) a far lesser man, succeeded to the leadership of the clan. Katsuyori's allegedly despotic ways, whereby he refused to take the advice of the old generals, led to defeat at Takatenjin (1574) in Tōtōmi province, and the fatal reverse at Nagashino in Mikawa province the following year. Matters had by then reached such a pass that it is thought that Katsuyori's headstrong manner coupled with the poor advice of two favourite councillors, caused his father's remaining generals to despair for the future of the clan and led them to virtually commit 'collective suicide' at Nagashino.[15] Katsuyori transgressed all three of Sun Tzu's aphorisms, which are:

III, 20   When ignorant that the enemy should not advance, to order an advance or ignorant that it should not retire, to order a retirement.
     22   When ignorant of military affairs, to participate in their administration. This causes the officers to be perplexed.
     21   When ignorant of command problems to share in the exercise of responsibilities. This engenders doubts in the mind of the officers.

The ko-ryū masters appear to have taken to heart Sun Tzu's stark warnings and encapsulated into the structure and methodology of their separate traditions' awareness a thorough understanding based on 'Know the enemy and know yourself; in a hundred battles you will never be in peril.'[16] The composition of these traditions in successive levels was designed to increase and reinforce the student's confidence. Each series of forms, the component parts of which, if taken separately, are as difficult as any earlier ones, bolstered his confidence as they seemed, over a long period of time, to be easier to absorb. In the Hasegawa Eishin-ryū, as our model example, the eleven forms of the Omori-ryū, while considered to be basic and therefore easily mastered, are anything but easy; they are not intended in the main to be practical fighting techniques but exercises leading towards the heihō forms. They take a long period of time to assimilate and master, usually considered to be as long as eight or ten years. When the student reaches a reasonable degree of understanding in these basic forms and at least part of the principles they contain, he is faced with the next level composed of older forms. By now his mind is broadened and the technical aspects of the new forms are easier to absorb. Few of the medieval daimyō, beset with all the problems of ruling their domains and the convoluted politics of survival, could afford such a long gestation period devoted to mastering the complexities of warfare. As in ancient Chinese times, they faced daily problems, large and small, and turned to recruiting experts in this vital field in the hope of finding those who would 'preserve their house and extend their lands'.

We do know that by the very end of the sixteenth century, Kenjutsu, now metamorphosing into *gekken* as the Edo period began to mould the structure of the warrior class to suit the Bakufu's aims, was formulated in three successive layers. The lowest, and largest, was instruction in swordsmanship for the ordinary samurai and this eventually evolved into modern Kendō; the second level combined the general training with more detailed instruction in the art of war so that these middle ranking retainers were able to function in command should they be required to go to war. The third and upper level, comprised the middle layer together with specific instruction in heihō. These three layers were still to be discerned in traditional Kendō to modern times. My own studies and conversations with very senior masters in Japan indicate that these divisions might date back at least to the middle of the Muromachi period, the fourteenth century, and most probably much earlier.

It would follow that the Sengoku daimyō, if they were sensible and wise, would ensure that they were at least aware of the principles of strategy and that their generals understood the same. It was men like Yamamoto Kansuke and Tsukahara Bokuden who were able to give this instruction despite their relatively ordinary warrior rank.

■ CHAPTER 7

Foreknowledge

When you are ignorant of the enemy but know yourself, your chances of winning or losing are equal.

When you are ignorant of both your enemy and yourself, you are certain in every battle to be in peril.

This is a key teaching of Sun Tzu but few understand its true significance for a number of years. Every new student of the traditional martial 'arts', and I use the term 'arts' advisedly, has been told in one way or another that until he masters the principles and can apply them, his chances of survival, in real terms, should he be involved in combat are, at the most optimistic, 2 : 1 against. That is if the opponent is exactly on a par with him in ability. Where there is a disparity because the opponent has greater experience, the chances of survival are far, far less.

Both the medieval Japanese rulers and the Chinese before them made extensive use of spies and agents. Such an organization, at every level, was of utmost importance to these rulers. Valuable information could come in any number of forms, the most reliable of which were trained observers who, from their experience, noticed details most others might miss. When armies formed up for battle, the experienced officers would seek to recognize the devices displayed on the enemy banners. In static warfare in all periods, the taking of prisoners in raids had the objective of finding out their units and a host of other details that we have mentioned before. Both these types of information gathering have been commonplace in war for millennia and have changed little since the time of the Chinese military philosophers long ago. However, this is very low-level intelligence and of short- or medium-term value; the general, and the title comes from the original meaning of 'strategist', needs to gather much deeper information far ahead of possible hostilities. He needs to have agents actively reporting back from every neighbouring state, friend or foe. He needs to know everything, civil or military, and however apparently insignificant.

When the founders of the major bugei traditions formulated their systems, they reduced the principles to their barest essentials. At the beginning of his fifth chapter, Sun Tzu points out that 'the management of the many is the same as the management of the few'; it is simply a matter of application. The forms of these classical ryū only obliquely address the gaining of 'long-distance' advance information; the masters accepted that this requirement would already be in place. They concerned themselves chiefly with 'shaping' the opponent in the short term and then acting. While they may not have stated the long-term need for intelligence, nevertheless the student was advised to observe his opponent carefully long before the fight was joined.

This early stage involved assessing his demeanour, attitude, the opponent's state of mind, and other factors as he approached combat so that the master could form accurate judgements well in advance. Conversely, the master behaved as naturally as possible so that he gave no similar information to his opponent. These considerations had the effect of creating a whole set of formal preliminary moves intended to mask the internal condition of the participants; the more natural one's behaviour, the more difficult it was to be 'shaped'. In the bugei these formal moves were often coloured by religious ritual, the latter being applied to the former in the same manner as the opening and closing rituals of 'normal' practice but with an added importance.

It is necessary to remember that we are not looking at these matters from the standpoint of the present day. These masters of the fifteenth and sixteenth century all came to the bugei with a thorough understanding and experience of war. They were well aware that war was a matter of life and death.

In the dōjō setting, and one surmises that little of substance has changed in the last five-and-a-half centuries, the formal examination and execution of these disciplines – both before and after practice – are maintained in their appropriate places, the beginning and end. The succeeding forms, themselves, are not so cluttered but focus on the short term. In these classical entities all superfluous material has been pared away. Simplification is always paramount.[1]

Shih

The heihō, therefore, depend on foreknowledge and this foreknowledge, in its most simplified form, requires repeated practice so that gradually realization of the 'essence', the actual principles involved, germinate in the memory (some would prefer to describe this as the spirit) and the student advances to the stage where he can apply the teaching. To reach this stage the deshi must free his mind and he must be able to improvise and use his imagination. The application of imagination equates to flexibility; it can also be expressed as *intuitive* action. In Iai-jutsu, the lowest common denominator of regular Bujutsu study, one man initiating the attack or threat and the other dealing with it, is reduced yet again. In this close-space area of swordsmanship the rôle of *uchidachi* is eliminated in the physical sense and the swordsman becomes *shidachi*, but the action depends on the application of the theory against a tangible opponent or opponents. Uchidachi has to be visualized by the swordsman.

The Iai deshi literally has 'to see ghosts'. Without developing the ability to fully create an image of his opponent(s) he will fail in his objective of mastering the principles. It is not only necessary for the swordsman to develop his ability to 'see ghosts' but also to achieve a level in so doing where he not only creates this phantom but *cannot control it*. Now, it must be understood that this is not an idle claim but advanced in all seriousness. It is a phenomenon closely akin to pure 'theatre' where an actor is able, through his developed art, to project to his audience his total absorption into his assumed character. Here we have one of the several aspects of Iai-jutsu that the student gradually progresses towards and achieves.

Once he can 'see ghosts' then the deshi begins to understand the structure of each form and the validity of the moves; the succeeding stage will take him to intuitive understanding of the fundamental principles of fluid movement. He will achieve the authority in his movements that will create 'confusion' in his enemies' minds. Put in another way, the swordsman will create in his own mind the images of his enemies, but those phantoms will be outside his control. From this point they will, in his mind, assume an independent 'life' and he will be unable to control them. He must learn to act exactly as though they were real opponents and destroy them exactly as though they were living entities to his own certain satisfaction. Anything less than this conviction of victory and, metaphorically, he has failed. Failure, in reality, is total and irrevocable; it is death.

There is nothing in this *shih* that is esoteric; this is not an 'altered state of mind', a hallucination, but a total immersion of self. When some people have the good fortune to see such a great master in action, so to speak, it is frequently reported that for a moment they, themselves, thought that they actually 'saw' his opponent, such is the power of the swordsman's conviction and his ability to 'project' his ghostly adversary.

This raises an interesting question: Iai-jutsu kata, like all the bujutsu forms, depends largely on predetermined movement. This is, after all, the only way that a warrior using 'live' weapons, especially swords, can at close interval safely train himself for combat. It is, in essence, military drill and as such has characterized all martial preparation for warfare since man first organized the use of group force. The Iai student by using his developing powers of visualizing his enemy cannot, as we have pointed out, actually control him. At once he introduces an element of unpredictability and yet he must deal effectively with the situation as if it were one of life and death.

Every master of these important ko-ryū that I have talked to in Japan has stressed that there is almost nothing predictable in warfare. They constantly paraphrase the ancient authorities as did Mao Tse-tung. 'In the business of war,' wrote Sun Pin, 'there is no movement that can be relied upon at all times.' Sun Tzu wrote that these variables are 'as inexhaustible as great rivers'. In the ebb and flow of battle, 'the fight can end at any moment'.[2] In the Kashima Shintō-ryū, extant from around 1525 or 1530 but based on much earlier teachings, this fluidity and flexibility is greatly emphasized in the teaching method. There is a constant juxtaposition of forwards and backwards movement in both 'attack' and 'defence', moves to the left and the right, and sudden vertical dropping movements. Almost as soon as the student has absorbed the ordered structure of each form, so the master increases the tempo of training or as suddenly slows it down. If the present-day method reflects the varied tempo of the late-Muromachi training, and there is little reason to doubt it, then the novice – albeit probably an already experienced swordsman – is given no time to think, only to act. This happens in Iai-jutsu too, but in the Kenjutsu there is always the added chance of the master (*uchidachi*) injecting a surprise move without any warning. Each situation addressed by the kata forms must be approached as though for the first time. The attitude of mind is vital; alertness must be constantly maintained so that any unforeseen eventuality can be instantly dealt with.

Reflecting on the experience of the medieval methodology in heihō, one is constantly amazed with the rapidity of progress from what is at first relatively easy, largely because the tempo is kept reasonably slow, to movement akin to greased lightning in a very short space of time; days rather than weeks or months.

Naturalness of Movement

The key to the physical mastery of bujutsu lies in the postures that are employed and how these postures act as a 'springboard' for the action that follows. The tora-no-maki extant today frequently contain crude 'thumbnail' sketches illustrating or accompanying named forms. Visually, they are the forerunners of the 'strip cartoons' that until recently helped to teach sporting excellence in the fields of tennis, soccer and golf, amongst many others. Nowadays, even these economical line drawings are replaced by photographs and computer-generated animation. The medieval master used these simplified graphic presentations to jog the memory of his deshi. Like Confucius, the master showed his student the beginning of the theory but expected him to come back with the rest, and by so doing, revealed nothing of the inner secret to the outsider.

These early line drawings, economically done with brush strokes, often contain a liveliness of movement, the elasticity of which suggests something of a coiled spring just waiting to be released. Whilst these are undoubtedly specialist representations they are also 'order-of-battle' postures. Sometimes their purpose may seem clear but usually this is not the case and they suggest in visual terms something of the unexpected. Unless the deshi is familiar with these myriad pre-engagement forms he may make incorrect assumptions and face a sudden bloody death.

It is interesting that many modern commentators who discuss the medieval iconography often severely handicap themselves by taking an approach that only sees the immediate physical representation but not the underlying intention of the artist or the commissioner of the drawing. These are *not* sketches meant to reveal some technique or other. They are as carefully conceived as the title given to the form and intended to present the *initiated* student, as he advances and deepens his understanding, an 'invitation for the commander to imaginatively shape the metaphor to suit the unique circumstances and thus use its poetic suggestiveness to his own ends'.[3] All too often the modern commentator makes observations which sadly miss the true point completely, simply because he is evaluating these graphic *aides-memoire* in the same way as the total novice or with only the barest familiarity with the actual entity. He is seeing only the surface and is unaware of what lies beneath.

When the master swordsman, the fully trained military expert, saw an unfamiliar technique whilst observing a *katakiuchi*, or live-blade 'duel', for example, he very quickly saw 'inside' both the participants' minds. He may not have been precisely clear as to the mechanical details but he easily understood the principles of the tactics involved. This holds as true today as it did in the late-Muromachi period. His intuitive understanding was based on sure knowledge and it is this that made the unattached master so valuable an 'agent' were he later to be employed as a strategist.

Such drawings as those we see originating in the Sengoku-era Japan or as far distant geographically as the now famous *Fechtbuchs* of Hans Talhoffer, produced in Germany between 1441 and 1467,[4] amongst several others, *may* show us the beginning of each 'form', it is much more likely that this is not the case at all and they present us with a synthesis of the master's teaching that only someone with many years insight will understand. This insight is, of course, *zanshin-no-ri*, or 'inner awareness'. The devil is in the detail, not the surface, and it would be a great mistake not to recognize this and here we have the error made by many modern commentators. They simply lack the experience that will afford them the insight.

Lau and Ames discuss the meaning of the Chinese character *chih*, 'to know', and point out that this is understood as 'situational'. 'To know and to act are one and the same'; we return to this aphorism time and again simply because it lies at the core of heihō. These drawings in the *ō-tora-no-maki* or *denshō*, as well as those produced in fifteenth-century Europe, were documents that, if created today, would be marked at the very least as 'Secret' but probably given the highest classification. They were as secret as the techniques themselves and were produced solely to stimulate the imagination, the 'inner awareness', of the recipient, Even then, unlike modern manuals, they absolutely avoided any detailed written description, leaving only the title of a particular form or the briefest outline of its meaning as a suggested image and the drawing as a visual stimulus.

In any case, few of these drawings that are now so freely available to the laity, depict, even if we would like them to do so, the actual beginning of a technique. Most come in, at the very least, a movement or posture or two later. Perhaps this is a point worth bearing in mind before venturing further into the unknown?

Hopefully, the above helps to clarify the situation, which brings us to consider what precisely these illustrations tell us. First it must be remembered that they were originally drawn at a very great distance of time before the present. The entities they represent may have subsequently evolved and developed through a number of stages since the medieval period, and are possibly radically different to the understanding that we possess now. By the same token, the subtle changes that may have entered over the last five hundred or more years had *not* taken place when these illustrations were drawn which means that we are effectively divorced from their original meaning. If we base our interpretations on the first point, we may be closer to an accurate assessment, but if based on the second, then great reservations need to be expressed. Both sets of interpretations will contain flaws, in our opinion, and are open to argument; the first less than the second, of course. Sun Tzu pointed to this when he wrote: 'The primary colours are only five in number but their combinations are so infinite that one cannot visualize them all.'[5] Apt comment!

The amount of information that may be extracted from the illustrated *tora-no-maki* and contemporary *gunki-emaki-mono*, illustrated war scrolls, may be limited to the detail in the drawings. Early denshō tend to display very cursive and often loosely brushed figures, quite small in size. They cannot be expected to convey much to the uninitiated other than in a general way. One of the most interesting things is how closely the very deep splay-footed postures drawn in these scrolls match the *kamae* employed in the same ryū at the present time. An important detail needed to evaluate these Sengoku period drawings is the precise handgrips used to grasp weapons. Unfortunately, these are often only treated in a cursory manner, too, not being considered important to the initiate. Another difficulty concerns the representation of the weapons flourished by these figures.

Polearms, like naginata or yari, are reasonably recognizable but care is needed where long swords are displayed. There seems to be very little distinction made between the long *tachi* slung swords and the *ō-dachi* or *no-dachi* long-hilted weapons favoured by some warriors during the mid- and late-Muromachi. In the case of both the last named there is sparse knowledge about their actual use and no contemporary traditions appear to have survived to guide us, only a very few vestiges here and there. A further complication, not answered because of the lack of precise detail, is the fact that it is known that a number of kenshi, Tsukahara Bokuden for one, used what is described as a *chōtō* in some of his live-blade *shinken-shobu*, the accounts not telling us the exact nature of this weapon. It is, in fact, thought to be a tachi or katana with an elongated hilt, a type of weapon termed a *nagatsuka-no-katana*, or possibly an alternative name for a short-shafted naginata or nagamaki. Already the reader will appreciate the minefield that the commentator has entered.

The illustrated tora-no-maki dating from the end of the Sengoku period and into the first fifty or sixty years of the seventeenth century were produced at a time when some of the best known swordmasters were enjoying generous patronage and so, it would appear, could commission professional artists to illustrate these scrolls.[6] Very good examples of such work are contained in the denshō of the Shinkage-ryū and the Taisha-ryū. These drawings, and there are many others from around the same era when bujutsu was changing to budō, that depict excellent details of, for example, foot positions and hand grips, do not necessarily have the same objective as the denshō of a century before.

When evaluating these illustrated denshō we must return to the 'poetic imagery' of each form's title and here, again, we run into that 'barrier of obscurity' where the name might seem to imply something but tells us little of any substance. Referring back to the three pictorial *fechtbuch* manuscripts of Hans Talhoffer, the first two may have been drawn by the master himself, but the third and best known, dated 1467, is the work of a skilled artist. Apart from a few drawings showing postures, at first glance the bulk generally appear to be single two-figure illustrations; then one realizes that many of them depict action from two separated parts of a 'mini' passage of arms – a 'kata' form in Japanese terms. Viewed in this way they illustrate a salient point that Talhoffer considered important within the 'form' and thus focused on its underlying meaning. Entertaining as they are to those purporting to 'devise' medieval battle re-enactments, neither these European drawings nor the Japanese denshō depict the start of a technique but a point within the teaching that will stimulate the student at some future point to intuitively realize a possibility that may have been staring him in the face for years and he had not realized the truth.

It is a common saying, now sadly largely ignored, that 'a little knowledge is a dangerous thing'. This is especially true in the classical bujutsu where interpretation must be both innovative and intuitive. Gradually, the student acquires the foreknowledge that is firmly based on the progression of the teaching from the simple and easily understood to the more complex and difficult. A metaphor that is often used is to liken understanding to the growth of a small tree from a seed to a small sapling putting out its first pair of branches, then a year or two later, those same relatively simple branches flower and bear the first fruit. After a long period, the now increasingly imposing tree is a complex structure spreading out wide and skywards. Some twenty-five years ago, I was instructed in one of the forms in the upper level of the Hasegawa Eishin-ryū by my master, who passed away three years later. I practised this particular form, the underlying principle of which I well understood to be based on fluid deception, but increasingly felt that there was something else there that had somehow eluded me. Then, one early morning about fifteen years later, I was running through this Iai-jutsu series in my mind before leaving to take *kangeiko*, dawn practice, when I realized what was missing. Just before his demise, at almost my last stay at my master's dōjō, he had shown me this form yet again, not just once but several times, but he had demonstrated it 'mirror image' – and I hadn't realized the fact!

Maybe this was my moment of *reimu*? The truth was there all along, within my grasp, but I had not seen it. Foreknowledge is not just a matter of mastering and perfecting physical movements and the theoretical concepts before applying them to a situation in a more and more realistic or efficient way; it is the intuitive understanding that can harness the unpredictable and turn it to an advantage. The sudden realization, without even a sword in hand, gave me an insight into the 'regular' technique, the orthodox 'cheng' form, in itself 'ch'i', that changed it to the unorthodox 'ch'i' that it really contained. At once, both forms seemed to extend my understanding far beyond their hitherto limited confines.

Returning to the images that illustrate some of the oldest denshō or 'tiger-scrolls', whilst the postures may often seem extremely crouched, indeed the figures' thighs are frequently almost horizontal to the ground, they have a common feeling of contained flexibility, an innate suppleness that presages the devastating explosive actions to follow.[7] The denshō of the Kashima Shintō-ryū and the Katori Shintō-ryū, in particular, illustrate this natural characteristic. The famous line drawings in the slightly later Shinkage-ryū, where several of the forms are contained in the *Tengu-shō* levels, have a similar dynamism. From the point of view of instruction, such deep 'sprung' postures take a very long time to acquire but once they are even partially mastered the forms begin to 'live'.

It seems possible that this emphasis on natural movement was developed from the deep influence of Tao within the classical heihō. We have already noted this in the swirling circular movements of the body and the sword that are intended to deceive the opponent. These deep postures drawn in the denshō both confuse and pressure the enemy since while they clearly contain an implied threat, they cannot easily be 'shaped'. They are 'orders-of-battle' without proper definable form. These Japanese postures of the medieval heihō entirely conform to the ancient Chinese view that warfare was (and is) an art and not a science. Lau and Ames commenting on the term *ping-fa*, explain thus: 'While an understanding of military operations will win battles, the long-term survival of the state requires a degree of understanding that goes considerably beyond battle acumen and skill.'[8]

Tao, or *Dō*, means the '*Way*'. It seeks to harness and harmonize man's actions in relation to nature. The classical bujutsu masters sought models from nature, understanding that naturalness of movement when fully absorbed into these arts led directly to effectiveness. They also understood that such natural images that they encountered in Sun Tzu such as 'water always finding its own level' conformed to the *I-ching*, the *Book of Changes*, that water was, therefore, 'formless'. The postures of the bujutsu denshō are like water – fluid, swirling, without shape. When coupled with weapons they are also fathomless, suggesting constant change from *yin* to *yang*, of *in* to *yō*, and back again. Therefore, to evaluate a posture one needs to fully understand not so much its physical form but its inner meaning. Studying Talhoffer's 'forms' in direct parallel to those of the Muromachi denshō shows that this truth extends to both. What appears on the surface is not 'fixed' but fluid.These postures hint at, but do not actually say, what is to follow.

To take one example from many; the shoulder posture known in Kendō as *hassō-no-kamae* is often termed *in-no-kamae* in the older bujutsu traditions. On the surface *hassō* implies a 'concept' or 'expression', but it is usually defined in modern swordsmanship as meaning 'many' or 'myriad'. This implies that it is 'a form without form' in that it does not suggest a clear reading to the opponent. A number of basic cuts can be made from this posture to add to the difficulty. It is also explained, though generally to relatively low-ranking students in order to fire their imaginations, that it is a posture assumed with a drawn sword whilst riding towards the enemy. The sword held thus at the right shoulder will not snag the *shikoro*, or neck-guard, attached to the helmet. All well and good and perfectly reasonable, but when we take the second term for the nearly identical kamae we at once encounter the inclusion of *in*, or 'shaded'. Now we have contained in the title of the form the suggestion of 'ch'i'. Nothing other than an implied threat of action can be read into the form; what that action might entail covers many possibilities but the enemy will find it virtually impossible to prepare a defence and it is equally difficult to launch an effective counter-attack. The subtle difference between the *hassō* and the *in* positions is that in the former the swordsman holds the sword with the *tsuba*, guard, opposite his mouth whereas in the latter the sword is grasped more firmly with the tsuba slightly higher and the right elbow a little raised. The dynamics of use from this point are quite different, the former 'hiding' the swordsman's intent whilst he responds to his attacker's tactics against the posture; the latter threatening an instant cut whatever the enemy decides.

It may be noticeable that in this discussion we do not look at group action, only the practice in the singular. The postures depicted in the denshō are intended to be fleeting; they can be both attacking and defensive; they are both orthodox and unorthodox. They cannot be categorized because of their inner formlessness.

In *and* Yō

Dualism is a constant theme in the Chinese and Japanese military writings from the most ancient times to Mao Tse-tung and the present day. It is the constant underlying theme running through the Japanese heihō in matters of the greatest importance in medieval warfare and hardly less simplified in modern budō. As we have seen, 'in' contains the meaning of 'shaded', deriving directly from 'yin', and can be considered as the 'negative' principle or 'the darkness before the dawn'. 'Yō', on the other hand, has a positive connotation; it is the 'first light heralding the dawn' or, etymologically, deriving from 'yang', the 'lit side of the mountain'.

From this, we can understand that the two are inseparable, the one passing into the other and back again as night follows day or as the 'moon waxes and wanes'.[9] Fluid contrasts are constantly encountered within bujutsu and later budō methodology, exemplified by the undulating tempos contained in the training.

While a great deal has been written on this concept it is interesting to examine the more basic expression to be found at the oldest extant practical levels. As might be expected, in swordsmanship we can find both 'in' and 'yō' expressed as kamae, postures. The *in-no-kamae* frequently occurs in the simple drawings contained in the tora-no-maki or denshō. We understand that the posture contains a wide range of attacking possibilities, none of which can be properly 'shaped' by the opponent. Less frequently encountered is the *yō-no-kamae* which has a pedigree passing down to the Ōno-ha Ittō-ryū (early seventeenth century) from the Kanemaki-ryū (second-half of the sixteenth century), itself deriving from the mid-Muromachi period tradition, the Chujō-ryū-no-heihō (fifteenth to sixteenth century). These two kamae represent the concepts, therefore, of 'shade' and 'light'. However, to complete the metaphor we should also seek the exact moment between the 'darkness before dawn' and the 'coming of light' within the heihō. Sure enough, an anonymous master swordsman turned to esoteric Buddhism for inspiration and from the Shingon mysticism devised a posture which is known as 'kongō-no-kamae'.[10]

Before describing the purpose of this kamae we should consider, in broad terms, the three major influences that eventually formed the basis of the Japanese heihō. Clearly, the first of these originated in the Zhou period in ancient China and reached its most developed form in the melting pot of the Warring States around the fourth century BCE. It was the teachings of Sun Tzu that became pre-eminent and it is evident that this master may have totally eschewed any beliefs that asserted the deities or spirits could exert their influence on the art of warfare. In his final chapter on the use of spies, he wrote: 'Foreknowledge cannot be elicited from spirits or gods, not by analogy with experience, nor by astrologic calculations. It must be obtained from men who know the enemy's situation.'[11] General Tao Hanzhang, a modern military authority, draws the above conclusion but perhaps Sun Tzu was subject to the growing Confucian beliefs taking root at his time?

Familiarity with the military classics spread through northern China after the Warring States period and may well have become familiar to the fierce nomadic Hsung-nu leaders during their protracted struggles with the Ch'in and then the Han. Taoist concepts centred on yin and yang attracted military philosophers as much as they did other scholars and one feels that the migrating Puyŏ tribes, during the course of their movements east and south into Korea and eventually part of their might crossing to annexe lands in Japan, may have been amongst the first to begin the process in the archipelago. It is clear that Puyŏ-Kayan military thinking was heavily influenced by shamanic beliefs originating in the Trans-Baikal region that stretched from the Manchurian plains west to the Altai Mountains and far beyond.

As Taoist beliefs began to manifest themselves they became mixed with the fast-growing cult of the female war-deity, Mārīcī, rooting successively in northern China and then reaching Japan during the early-Heian period. The Taoist esoteric practices, mixing with Yayoi and earlier native animistic beliefs, became established in the 'mountain' religion, known to modern scholars as *sankaku-shinko*, certainly by the middle of the Yamato period, say the sixth century, soon emerging with the proto-yamabushi who later formed the Shugendō sect.

The final influence that had a significant bearing on Japanese military thinking came early in the ninth century when the monk, Kōbō Daishi, established the form of esoteric Buddhism known as the Shingon-shū. As the recognizable *buke* class emerged during the Heian, so the beliefs in the esoteric deities came together and proved very attractive to the warrior groups. Even at this early stage, it was Sun Tzu who provided the theory underlying the craft of warfare whilst Marishi-ten (Mārīcī) in Shugendō and the all-powerful Buddhist 'Heavenly Guardian', Fudō-myō-ō, in Shingon, offered divine protection from enemies. In Japan, the esoteric practices, characterized by periods of extreme ascesis, became part and parcel of the bugei and the deeper philosophy of heihō.

In Shingon, the second fundamental sutra after the Dainichi-kyō is the Kongōchō-kyō. *Kongō* means 'adamantine' or 'imperishable'; *chō* means 'peak' or 'summit', or the 'crown of the head'.[12] In the religious sense it refers to Dainichi Nyōrai's unsurpassed wisdom. *Kongō-chō* comes from the Sanskrit *Vajra*, a thunderbolt or trident, and the Chinese *Chin-kang-ch'u*, a very ancient symbol indeed that can be traced back to the earliest Mesopotamian civilizations. In Buddhism it soon came to symbolize the victorious power of Knowledge over Illusion and evil influences.[13] From this teaching it is easy to understand the appeal that the *vajra* had on those bushi able to immerse themselves in the mikkyō.

While Yamada Jirokichi suggested that the roots of the Ōno-ha Ittō-ryū were to be found in the Chujō-ryū, and of course this assertion is true, there is something inherently archaic about this form where the sword is held upright in front of the face in the centre line. As always in swordsmanship, the 'hand of power' is the left. This at once harmonizes with the important *mudrā*, the *chi-ken-in*, which in Tantric Buddhism emphasizes the active male principle in the aspect of Knowledge. It is believed, therefore, to have the power to destroy the passions (evil) of the world. The strong religious connection to the mikkyō raises the possibility that this form goes back for its inspiration at least to the Heian period and is very old indeed.

Apart from the double-handed kongō-no-kamae of the Ōno-ha group of ryū-ha the only other instance of its use that the author is familiar with is the final fleeting posture with the sword in the Shiten-ryū Kiome-no-waza kata of Iai-jutsu where the sword passes from the two-handed to the one-handed form in-no-kamae as a symbolic adamantine warning to the exorcised spirits dispersed by the swordsman.[14] In this case, the posture suggests, through the use of both hands, the Oneness of Knowledge and Principle; the application remains the adamantine firmness in the swordsman's soul.

Here, then, taking just these three postures, we can sense the fusion of the concepts of yin and yang from the Chinese roots, with the additional power conferred by the much later development based on Buddhist esotericism, added, at least partially, to bestow authority on the swordsman's inner zanshin. The orthodox and the unorthodox are united precisely where the two aspects of light and shade meet, yet this fusion can only be expressed in terms of metaphysical symbolism. This is not, one feels, a fighting posture that can be expanded and applied to the battleground, yet it symbolizes that Knowledge and Principle firmly rest in the commander and affirms that he is able to combine the ch'i and cheng as he wills.

Summary

As we have noted, Sun Tzu may have been a materialist with regard to his beliefs concerning the influence of spirits and deities on warfare, but it seems that the advent of Taoism and the later esoteric

Buddhism soon caused lesser warriors to hedge their bets, particularly where the promise of metaphysical 'protection' could be bestowed. By following certain rituals of empowerment and using the appropriate formulae, the spirits and gods could be enjoined to assist. The spirit world was too real, too close, and too appealing to be ignored. Typically, the Japanese warrior steered the pragmatic course by following Sun Tzu but also invoking the deities to help. By the Muromachi period Marishi-ten reigned supreme. The practical advice of the Chinese masters flourished alongside the metaphysical rites.

Ritual had an important modifying rôle in the development of the heihō but the Japanese characteristic of practicality also asserted itself; an example of which can be seen in the generally rational and efficient design of bladed weapons throughout the medieval period, though with the exception in some cases of those created in the Edo period. The difference between such weapons from Japan compared to China and other parts of eastern Asia needs no further comment.

Kamae

We can understand from the addition of the kongō-no-kamae to the concepts of *in* and *yō*, the Japanese development of the *yin-yang* principle. The Japanese do not engage with the Western ideas of black and white, preferring, on the whole, shades of intermediate grey. In-no-kamae and yō-no-kamae are the two extremes and the medieval kenshi understood the meanings very well. In reducing the ideas to the rationale of heihō, the inclusion of such a linking posture would have had some appeal.

When we focus more specifically on Iai-jutsu to explain and illustrate the complexity of ch'i movements, we see immediately that the kamae of swordsmanship and the various types of polearms are clearly simplified 'orders-of-battle' representing the positioning of military formations, large or small, on the battlefield prior to the commencement of hostilities. In the closer interval Iai not one of these positions appear prior to the outset. It is a famous saying in these 'Drawing-sword Arts' (*Battō-jutsu*) that 'Iai is in the scabbard'. As always, the bugeisha is fond of pithy terse maxims which, when analysed, contain a world of meanings. It should be expected that the medieval weaponed arts will represent at first the disposition of the warrior groups. It follows that at this initial stage such dispositioning can be reduced for bujutsu purposes to 'orthodox' postures designed to confuse the opponent, just as the unexpected movement and repositioning of flags and banners will confuse prior to a battle. Iai-jutsu, on the other hand, has a much greater immediacy. Any preliminary posturing would warn of intended action and, if such a warning is heeded, trigger counter-measures.

The use of different postures in Iai-jutsu is contained within the flow of action and so is, by definition, 'unorthodox'. The initial postures of bujutsu, on the other hand, are often assumed at a relatively long distance and sometimes maintained for a fairly long time prior to any subsequent action.

Returning to the premise that 'Iai is in the scabbard' it becomes apparent that this being the case there is no reason to have preliminary postures. One of the most difficult aspects of Iai-jutsu for many deshi is how to close the distance to come within striking range of the opponent *without* communicating to him that danger will inevitably follow once the swordsman is within reach. The kenshi must close this distance without warning by his 'body-language' of his intent. His movements must be as matter-of-fact and natural as they can be, but nonetheless he must reduce the interval to one within which he can operate. Perhaps this can be described as the use of 'negative-kamae'? The fact remains that in order to gain and keep the initiative, the swordsman must give no warning.

In military terms, the lone swordsman in the Iai-jutsu situation is a very weak force; one that if detected can be destroyed in a matter of seconds; in minutes in a larger situation. The guerrilla commander cannot deploy his forces, however large or small, in the orthodox manner. Mao defined guerrilla strategy as based primarily on alertness, mobility and attack. It must adjust to every situation. This definition exactly defines Iai-jutsu.

Kamae in this 'irregular' aspect of the bugei is contained within the movements that make up the forms; they are, in this sense, internal rather than external. They exist but are encompassed within the overall movement and are employed in concentration and dispersal.

Considerable teaching energy is expended on getting the modern Iai student to be aware of the importance of *jōdan-no-kamae* within the forms, a posture that usually comes between the attacking draw cut and the second decisive stroke. Since the jōdan has been a principal preliminary posture from the very beginning of two-handed swordsmanship, then one might reasonably suppose that in the mid-Muromachi it was examined in depth by the bugeisha.

'Alertness, mobility and attack' are the essential constituents of ch'i warfare but Mao states this as the barest definition. Like Sun Tzu he does not go much further into detail but leaves the interpretation to the commander on the ground. In our model of Iai-jutsu as an illustration of unorthodox tactics and strategy we have the initial approach or appraisal of the situation which is 'alertness' in Mao's definition. The second stage is the attacking cut and whatever may follow where the need is to deal with one or more opponents. This is the 'mobility' or 'attack' phase – necessarily short and explosive, but never protracted. It is these two stages that contain ch'i action and it is here we find the use of kamae presented as part of the 'coiling', fluid, formless strikes. However, kamae have an inherent purpose even if they are internalized.

The jōdan is no exception. Usually, it is linked to the Buddhist element 'fire' and, in bugei terms, means 'strength'. It is an offensive posture rather than a defensive one, unlike *seigan-no-kamae*,[15] the 'middle posture' or 'aiming at the eye' posture, which is neither one nor the other but contains elements of both. In Iai, after the 'draw-cut', the sword is generally whirled to the left, back past the left ear and up to jōdan where it is brought to the slightest pause above the head as the left hand takes hold of the *tsuka*, or hilt. This pause is 'concentration' and is noticeable in many Iai forms. It permits a 're-grouping' or focusing of the integral forces to re-assess the situation before the attack continues but, as importantly, it is an inner reminder of the need to 'act on sure knowledge' that is firmly based on well-tried principles.

Here, in the simplified model of close-space swordsmanship, we find addressed one of the problems of unorthodox actions, the avoidance of the dispersal of one's forces after the first strike. In the lower levels of Iai-jutsu, most forms predicate one-to-one encounters; it is in the upper levels that the theory is expanded, usually to two opponents but sometimes as many as four or even five. To maintain the initiative and regroup after each part of the attack means that the swordsman is always in control, alert and mobile. His very movement is his strength. Translated into a guerrilla war situation, and supposing that this were a night operation, the darkness can be a problem for both sides. For the guerrilla, inferior as he is numerically compared to the regular enemy force, he must make the darkness a strength. Being able to concentrate and then disperse is important. Understanding this principle underscores the value of the jōdan or other 'attacking' postures studied in the Iai.

CHAPTER 8

The Distinction in the Heihō between *Ch'i* and *Cheng*

Mao wrote: 'The strategy of guerrilla warfare . . . is constant activity and movement.'[1] All that we have seen thus far about Iai-jutsu confirms this premise. The swordsman is only strong whilst he maintains the initiative and his constant flowing movement; this form of combat does not permit the situation where he becomes static and defensive. As Mao emphasizes, in such warfare there is no such thing as a decisive battle, there is nothing comparable to the fixed passive defence of orthodox warfare.

There is one particular form in Iai that seems at first to examine the problem of a successful night attack after which the swordsman is pursued and forced to stand his ground. On the face of it this could be defined as a *ch'i* attack that must change to a *cheng* when the guerrilla group cannot escape. However, the form introduces an unexpected tactic which changes the final *cheng* back into a *ch'i*. It not only does this in the 'official' form but also in the alternative one. We shall examine this in detail later.

The form outlined here also reinforces the fluid links between the use of an unorthodox attack launched against an unwary enemy and an almost identical technique used in defence but made unorthodox by the addition of a completely new component of surprise. There is no 'chasm between the two' (*ch'i* and *cheng*).[2] Such differences as do exist may change to similarities as the circumstances alter. The forwards attacking 'draw-cut' of the *ch'i* becomes a rearwards stepping defensive cut on the withdrawal steps. It still makes use of a sudden and unexpected draw since the kenshi is aware that he is pursued (albeit in near darkness) but by either jumping up at the instant of drawing or dropping to a crouch at this moment gives the initiative back to his hands and the pursuer is destroyed. This form is one that even in its dōjō setting maintains its clear link to heihō in that the student must visualize the whole of the action taking place in near twilight or darkness and that his footwork across the floor is absolutely noiseless. Like several other forms in the Iai corpus, there is here a very strong sense of realism that is only removed from actual warfare by the need for secret training. The Takeda master-strategist, Yamamoto Kansuke, noted that 'when you are fighting in the dark, keep your body crouched down and observe you enemy's outline carefully to see what kind of weapon he has'.[3]

The Rationale

It seems to be a characteristic of some classical and all modern budō to seek technical explanations for every detailed aspect of each entity. Miyamoto Musashi as far back as the early seventeenth century levelled criticism at many of the ko-ryū that by making the forms that they taught, and presumably the rationale that went with them, so complex, they guaranteed the head instructors an income from every student for years to come. One wonders if indeed this was the case during the dynamic decades of development during the mid- and late-Muromachi?

One suspects that most masters practised first in order to gain the insight and expertise; it was only in their later years that they had the leisure time or inclination to devise the complex rationale. We know that many of these kenshi 'followed the yamabushi paths' but few modern commentators writing in Western languages seem to have seriously considered what this phrase might truly have meant.

THE DISTINCTION IN THE HEIHŌ BETWEEN *CH'I* AND *CHENG*

The suggestion that springs to mind that these were the methods of severe training allied closely to hardening the body and mind through physical and mental disciplines almost unimaginable to the modern student.

These kenshi undertook, voluntarily, ascetic training coupled with the practical wielding of weapons that must have been arduous in the extreme but had the aim of shortening the period of time required to gain inner understanding. Few were truly successful, we would deduce, or there would be many more ko-ryū recorded than we know of today. It was probably the immediate successors of these dynamic 'founders' who then created the polished teaching rationale that we are bogged down in today. Like Musashi, we may often feel that such a plethora of terminology is created solely to complicate matters, to preserve the upper levels, and have little or nothing to offer the heihō-jin who simply identifies the basic principles and applies them. Musashi, like Mao after him, expressed his understanding in simple terms, but both eventually 'dressed things up' when they had the time to write in detail.[4]

Perhaps we can identify the early stage in the development of the rationale as being that which looked towards the esoteric practices of both Shugendō and the Buddhist mikkyō. In the latter case there may have been a significant influence exerted by the Tendai-shū but most authorities in Japan are agreed that it was the Shingon-shū that was by far the strongest of the two. However, within the ko-bujutsu, opinion indicates that it was the methods and secretive teaching of the Shugendō yamabushi that was dominant; the problem in seeking the truth is that all three came together and overlapped at various times and that much of value was swept away and lost forever during the reforms of the Meiji period, changes that proved to be a cultural crime of enormous proportions. Shugendō alone is considered to have lost more than ninety per cent of its size and background, suffering the destruction of many its temples; Shintō was reformed, records lost, archives destroyed; all for the sake of a new 'conformity'. All this came in that period when the martial entities became *passé* between 1877 and around 1900. This might seem a short period of time but it was one of great changes following two major civil wars and the complete upheaval of Japanese society changing from the medieval to the modern; a period when ko-bujutsu and classical budō also lost, irrevocably, three-quarters or more of its body. It is small wonder that the shin-budō of the present day often cloud the issues with terms and definitions that nobody really understands or needs; it is a process that has been going on since the last echoes of the great conflicts of the Sengoku period finally died away.

■ CHAPTER 9

The Influence of the Mountain Religion

In the development of warrior understanding there seems to have always been a clear connection between physical skills, a search for intellectual understanding and a yearning, possibly unvoiced, to secure protection through the invocation of the deities. There is a marked leaning towards shamanic practices of summoning divine aid that were in the most remote period associated with bird dances and deer. These very ancient traits can be traced back to the mid-Yayoi period[1] and are suggested in mythological accounts in the oldest chronicles in connection with the first land-taking forays of the Puyŏ-Kayan chieftains from Korea.

Awareness of Sun Tzu and Sun Pin may have first reached the archipelago at the latest with the final Puyŏ-Kayan settlers who flooded in during the fifth and sixth centuries, the same period when the proto-yamabushi began to appear together with their clear connections with Inner-Asian shamanic practices and as the 'protectors' of the Taoist recluses who had taken to the same mountain fastnesses in many parts of Japan. A more general awareness of the supposed power of Marishi-ten and the Buddhist mikkyō may have surfaced at almost the same time and both were quickly absorbed into the burgeoning military ethos. Whilst these influences have been discussed elsewhere,[2] it is noteworthy that from the earliest extant records within ko-bujutsu there is the clearest connection with the Shugendō yamabushi and the associated cult of Marishi-ten. Springing directly from this comes the belief in the tengu as the 'messengers' of this powerful female war-deity. The inner principles of the highest level heihō have always been imparted by Marishi-ten by means of her 'messengers'.

Many centuries after the yamabushi-Marishi-ten connections manifested themselves, a number of characteristics developed by these entities began to be explained within the rationale and, as one might expect, the explanations themselves reflected the *yin* and *yang* dualism, giving rise in certain quarters to Miyamoto Musashi's criticism of 'unnecessary complication' for the purposes of expanding the income derived from giving instruction in these arts and ways by the professional masters. A century before Musashi died (1645), the great Yamamoto Kansuke had pointed to the chicanery of those who claimed mastery by wearing what they purported to be 'Confucian' apparel.[3]

These characteristics, so laboriously set out in the later bujutsu, are divided into light and shade, the outer and inner, and are generally known as the '*bu-no-jutsu*' and the '*bu-no-ri*'. They can be summarized in pairs as below:

Bu-no-jutsu

Kangae-to-kan-no-jutsu
Outward initiative response to circumstantial factors.

Sen-no-jutsu
Outward flexibility in response to threats.

Fudoshin-no-jutsu
Outward steadiness of spirit in response to threats.

Bu-no-ri

Kangae-to-kan-no-ri
Intuitive understanding in judging a situation.

Sen-no-ri
Inward ability to anticipate changes.

Fudoshin-no-ri
Steadfastness of mind.

Hara-no-jutsu
Resolution to maintain inner equilibrium.

Hara-no-ri
Ability to strengthen the *hara* and knowing physical movement stemming from this.

Mukamae-no-jutsu
The outward understanding of form or posture.

Mukamae-no-ri
The intuitive understanding of an opponent's posture and its potential.

Kiai-no-jutsu
The ability to produce meaningful kiai and show correct breathing.

Kiai-no-ri
The ability through breathing to calm both body and mind.

Aiki-no-jutsu
To project outward harmony.

Aiki-no-ri
To maintain total inner harmony of mind and body.

Go-ju-no-jutsu
To be able to change from strength to weakness and back.

Go-ju-no-ri
To be able to harmonize the reciprocal forces.

Zanshin-no-jutsu
To develop awareness in all circumstances.

Zanshin-no-ri
To possess inner awareness despite all pressures.[4]

When we examine these nine basic traits found within bujutsu we can understand that several are developments from the teachings of Sun Tzu couched in terms that derive from the mikkyō. *Fudoshin-no-ri*, for example, makes use of Fudō Myō-ō in his aspect of an adamantine Defender of the Law; one who, once set on a course of action, will not be deflected from his purpose. *Go-ju-no-jutsu* and its inner aspect, should be quite obvious; and *Zanshin-no-ri* has also been explained. The 'technicalization' or 'compartmentalization' of the structure suggests to me that during and after the Edo period many masters felt the need to associate their budō more and more closely with the teachings of Buddhism and to give these theories a 'Zen-like' flavour in order to add appeal and attract students. The complexity militates against the fundamental axiom that 'to know and to act are one and the same'.

Whilst this observation is by way of a generalization, it does reinforce the view that in the warrior arts as they were originally conceived, and in the upper levels in particular where there was a pressing necessity for the bugeisha to apply the principles of warfare in order that he could command in battle, the understanding was heavily coloured by the 'mountain beliefs' of *sankaku shinko*. Whilst these beliefs are clearly complex, the 'inner' feeling for them is simple and undemanding. When considered from the viewpoint of the suggested origins of the medieval Japanese understanding of heihō then one can easily discern that the three levels of military competence amongst the warrior ranks were based on the ancient structures brought so far from the Trans-Baikal steppe. The Yamato military 'corporation' was the Mononobe who took responsibility for the conduct of warfare. The upper level of the warrior group needed to understand the philosophy of war but this depended very much on the propitiation and support of the deities and, once this patronage was obtained, the transmission of the deity's teaching.

All preparation for war was preceded by ritual designed to gain the support of the deities and has essentially remained unchanged except in the detail, for at least a millennium-and-a-half. The warriors were convinced that they enjoyed the deities' protection. Wounds sustained in fighting and defeat came through human error rather than divine failings. Victory required the successful leader, be he chieftain or general, to show his gratitude to the divine patron and this took the form of appropriate rituals performed by the war leader, or by suitable offerings of fine armour and weapons to those temples and shrines sacred to the war deities. These were martial customs associated with the preparation for battle and for victory throughout the Euro-Asian landmass but in the Japanese archipelago we see these rituals and customs in a form that has remained unchanged in substance for well over a thousand years. The 'upper level' bugeisha in a number of the oldest bujutsu ryū customarily performed a number of special rituals before the deities at several Shintō sites, the precise nature of these duties hardly studied by Western specialists, but their antiquity is beyond doubt.[5]

The early heihō-sha followed the yamabushi ways, as we have often noted; this may have meant that the first secret esoteric teaching was sought from certain yamabushi ascetics well-versed in the weaponed arts who were also able to intercede with Marishi-ten or, as seems to have happened, with Fudō Myō-ō. In the Sengoku period these two deities, the one from the Buddhist pantheon and the other of very ancient Central Asian origin, were often indistinguishable in the warrior mind. The art of war may have been contained and developed in the teachings of the Chinese philosophers; the understanding of these teachings lay at the disposal of these two preeminent deities, especially Marishi-ten.

The late-twelfth- or early-thirteenth-century legend of how the youthful Minamoto-no-Yoshitsune received his warrior training from a dai-tengu prelate in the remote Sōjō-no-tani behind Kurama-dera, simply confirms that by this time at least the structures outlined here were clearly in place and, in all probability were already mature, if not old. From the early-eighth century most of the important Shintō shrines were already constructed but were sited on locations long associated with the particular deity or other spirits who were tangibly enshrined there. There were already strong martial cults centred on many places, Kashima and Katori for example, some native in character, others exhibiting signs of an Inner-Asian origin that must have entered the archipelago with the usurping Puyŏ-Kayan immigrants. All of these traditions and ancient chronicles help to confirm and support the above argument despite the paucity of written records.

None of these beliefs was too demanding intellectually but then, as now, they required the upper-ranking warrior to have a simple determination to succeed. The nature of the inner teachings necessitated retirement to remote areas where the inner secrets were sought through undergoing severe ascesis, polishing heart and mind with the most arduous training, an aspect of the concluding process that seems to have lain firmly within the aegis of the hardened yamabushi. These quasi-shamanic 'guides' may not have been quite the same as the more familiar later Shugendō yamabushi of the Edo period but a survival of the much earlier proto-yamabushi who may well have remained active to the end of the Muromachi period. If the required understanding did not manifest itself during the periods when the kenshi undertook such musha-shugyō then a period of secluded 'gestation' within the grounds of one or other of the noted 'warrior' shrines followed the wandering phase, and 'enlightenment' was sought through prayer and hard practice. At this stage we come very close to a form of religious ecstasy known as *sanrei*[6] where intuitive understanding is closely linked to significant physical and mental deprivation coupled with a profound faith in the power of the deity but, despite popular beliefs to the contrary, nowhere in the late-Muromachi does one find this final stage of seeking *reimu* associated with Zen-dō. So far as the bugei traditions are concerned, Zen hardly exists; the influences are nearly always attributed to Marishi-ten, Fudō Myō-ō, and, by extension, the tengu. Occasionally, at the beginning of the illustrated *tora-no-maki*, one finds depictions of Shintō deities, usually single male and female figures, but even these are drawn as an aspect of syncretism harking back to the earlier deities.

Of interest, purely from the Shugendō aspect, is that the deities connected to the martial process rarely seem to include the fierce but mysterious Zao-gongen whose cult, whilst still active, reached its zenith as long ago as the twelfth and thirteenth centuries, very much declining since. Zao-gongen is always depicted in the iconography and sculpture as a fiercely martial presence but, so far as I am aware, is rarely found in the bugei tradition. This reinforces the supposition of at least two co-existent streams within the yamabushi structure. The problem of separating them is made extremely difficult by the great secrecy inherent in the Shugendō movement ever since it first appeared.

The esoteric influences are further complicated by the fact that not only do we find the mikkyō that can be identified with Shugendō but a blend with the rituals followed by the Shingon and even those associated with the secretive mysteries of certain Shintō deities. These latter include rites that are linked to the brother deities,

1. Marume Kurando-no-suke, who later took the Buddhist name of Tessai, was born into the Sagara clan in south-western Kyushu, present day Kumamoto Prefecture, in 1540. (He may have been born seven years earlier in 1533.) He took part in his first battle at the age of sixteen and when he was nineteen entered the service of Amakusa Naodane where he trained in the bugei. He is credited with mastering no less than twenty-one other arts of the bugei, with the addition of calligraphy, flute (*fue*) playing, and the tea-ceremony (cha-no-yu). His grave is in Ichibu, Nishiki-machi, Kyūma-gun, Kumamoto-ken.

2. Memorials to members of the Yagyū family at the Hōtoku-ji, Yagyū-mura, in Nara-ken. Many of the Yagyū swordsmen are commemorated here. Yagyū Munenori, himself, the youngest son of Muneyoshi, (died 1606), was buried in the Kotokuji-dera, Edō.

3. Wooden painted statue of Yagyū Munenori, made in Kyōto and erected here seven years after the great swordsman's death in December 1645, at the age of seventy-three.

4. In May, Eiroku 9 (1566), Fujiwara Idzumo Ise-no-kami Nobutsuna, the founder of the Shinkage-ryū, presented these four scrolls to his successor, Yagyū Sōgen Sekishūsai. These have been carefully preserved by the Yagyū descendants. The scrolls are named, respectively: 'Swallow Flight', 'Seven Tachi', 'Three Studies' and 'Nine Points'.

5. Amongst the many illustrations are detailed drawings of swordsmen opposing each other and a number outlining instruction by tengu as 'messengers' of the female war deity, Marishi-ten. The illustration accompanies the first 'form', termed *Ikkyo Ryōdan*, of the 'San-gakuen-no-tachi'.

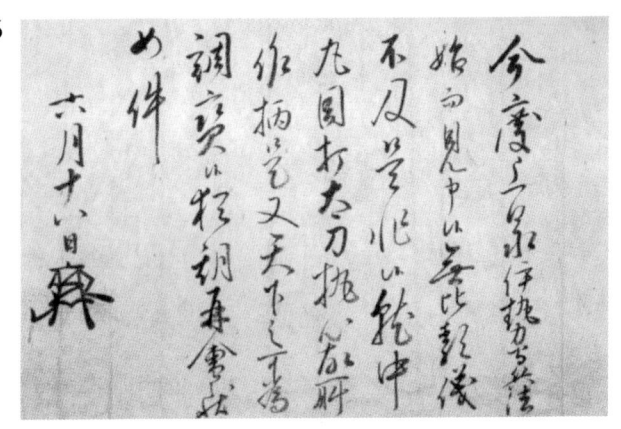

6. This is an extremely rare letter written by the thirteenth Shōgun, Ashikaga Yoshiteru, (1535-65), expressing his pleasure after witnessing the Shinkage-ryū-no-tachi skills demonstrated by Idzumo Ise-no-kami Nobutsuna and Marume Kurando-no-suke about the year Eiroku 7 (1564). The letter, preserved by the Marume family, reads:

> Today, for the first time, I witnessed the Shinkage-ryū of Ise-no-kami Uwaizumi. I recognize this as excellent bugei. Especially noted is the uchidachi mastery of Marume Kurando-no-suke which may be placed in a class of its own. I hope to witness this again.
> Yoshiteru, 18th June (?)

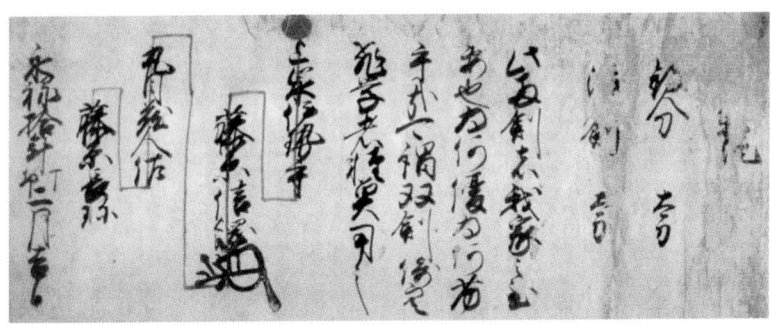

7. Licence (*menkyō*) presented to Marume Kurando-no-suke by Ise-no-kami Nobutsuna in Eiroku 10 (1567). At that time the tradition's name was variously written in different kanji which could read 'New-Shadow'-ryū, 'True-Shadow'-ryū, or 'God-Shadow'-ryū. The text approximately reads:

> Shinkage-ryū
>
> > Killing sword tachi
> > Let-live sword tachi
>
> > 'Both swords are indispensible to my house,
> > Which should be superior and which inferior?
> > As both swords are raised to the sky
> > Those who learn must never use them lightly.'
>
> Eiroku 10, February.

8

8. 9. Two illustrations from the Fuzeiken-Taisha-ryū *kumitachi* scroll. These Taisha-ryū *emakimono* were presented in Empō 3 (1675) to Obushi Kansuke by Kijima Nobumitsu Itō Yagōemon. The tengu-yamabushi illustrations contained in the scrolls remind the deshi of the powers of Nature, (earth, water, wind and fire) and whilst these are also Buddhist 'elements' they are also emphasized in Shugendō and the sankaku-shinkō, clearly deriving from Taoist roots and the cult of Marishi-ten.

9

10

11

10. 11. Tengu appear in many of the illustrated tora-no-maki, invariably taking the rôle of *uchidachi*, the 'attacker' in most weaponed ko-ryū kata. *Plate 10* depicts a particularly fierce and animated tengu, again in garments suggesting the yamabushi connection, opposed by a 'human' *shidachi* who wears *fukurō-hakama* and holds his sword in an unusual in-no-kamae. This is a ch'i posture. *Plate 11* is from the Shinkage-ryū tora-no-maki and depicts a tengu with a rather hirsute face named Chiraten(-bō). The form is called *Korandome*.

12. Iai-jutsu: A sudden evasive retreating move employing a turning draw and thrust against a pursuer followed by a devastating left-to-right *kesagiri*. The enemy does not realize that the sword has been drawn until it is too late. A similar ch'i tactic is contained in the first *okuden* level of the Hasegawa Eishin-ryū.

13. Iai-jutsu: This Iai makes use of strong sweeping attacks at two different angles, including dropping to half-kneeling. An excellent example of unorthodox ch'i tactics in action.

14

14. Iai-jutsu: Another ch'i tactic utilizing a turn towards an enemy to the rear accompanied by a devastating upwards *kesagiri* stroke and a final strong *kirioroshi* to complete the form. The sword is drawn in such a manner that the enemy is not fully aware of the danger before the *kenshi*'s blade sweeps up through his lower left hip; by then it is too late to defend.

15. A spectacular surprise vertical leap from half-kneeling whilst drawing sword. This leap is known as tengu-tobi. *Nukitsuke-no-ken* form, Iai-jutsu, Tenshin Shoden Katori Shintō-ryū (sixteenth century). Otake Risuke *Shihan*. (Photo: Author.)

16

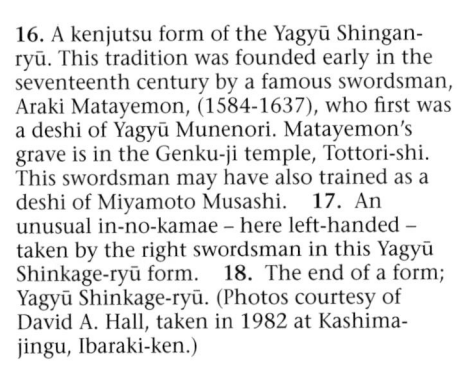

17

18

16. A kenjutsu form of the Yagyū Shingan-ryū. This tradition was founded early in the seventeenth century by a famous swordsman, Araki Matayemon, (1584-1637), who first was a deshi of Yagyū Munenori. Matayemon's grave is in the Genku-ji temple, Tottori-shi. This swordsman may have also trained as a deshi of Miyamoto Musashi. **17.** An unusual in-no-kamae – here left-handed – taken by the right swordsman in this Yagyū Shinkage-ryū form. **18.** The end of a form; Yagyū Shinkage-ryū. (Photos courtesy of David A. Hall, taken in 1982 at Kashima-jingu, Ibaraki-ken.)

19. 20. 21. Naginata-jutsu and kenjutsu demonstrated by masters of the Kashima Shintō-ryū at the embu (martial commemoration) at the Kashima-jingu in 1982 celebrating the 400th anniversary of the death of Tsukahara Bokuden, one of the most famous bugeisha of Japan. (Photos courtesy Phillip Jupp, Eikoku Kendō Renmei.)

22. 23. Kenjutsu of the Yagyū Shinkage-ryū. 24. A tengu-tobi leap from a half-kneeling posture in the iai-jutsu of the Tenshin Shōden Katori Shintō-ryū. (All photos courtesy David A. Hall, taken in 1982 at the Kashima-jingu, Ibaraki-ken.)

25. 26. Bō-jutsu (uchidachi) opposed by sword.

27. Kenjutsu. Note that shidachi (right) has assumed the very deep *'yubi-no-makura kamae'* often encountered in this tradition. This is a typical ch'i tactic. (Photos courtesy Phil Jupp, British Kendō Renmei, taken at Kashima-jingu, 1982.)

28. Various symbols mentioned in the text.   *Top left*: Shingan-ryū (founded late-sixteenth or early-seventeenth century): the yatagarasu and running hare associated with the sun and moon, symbolic of the teaching of Marishi-ten transmitted by means of the tengu.   *Top right*: The symbols of the sun and moon from a Shugendō honzon depicting many tengu.   *Centre left*: The nine barred Kuji-no-hō symbol of empowerment and protection often found in the bugei. Vocalized, the nine accompanying sounds in their proper order form an invocation to Marishi-ten   *Centre right*: Two mudā, symbolic of power and protection. Left: the *ongyō-in* Right: the *chi-ken-in*.   *Bottom left*: Two *mon* displaying the feathered fan of 'invisibility'.   *Bottom right*: Marishi-ten represented on two honzon as winged tengu in the garb of Fudō-myō-ō, each standing on the back of a fox. The intermixing of a number of Buddhist and Shugendō symbols from their respective mikkyō is inescapable.

29. Other cultic symbols that became associated with Marishi-ten (Mārīcī).
Top left: Sword koshira (hilt pommel) in gilt-bronze depicting intaglio the head of a griffin. Fifth/sixth century Kofun period, Okayama-ken.
Top right: Pierced iron tsuba from the mid-Muromachi period depicting two inoshishi (Marishi-ten) and characters with the name of Hachiman, the Shintō war deity. *Bottom left*: Marishi-ten, here represented by a yatagarasu within the sun disc and mounted on an inoshishi. Shrine honzon. *Bottom right*: Fierce winged tengu from Taisha-ryū tora-no-maki.

Takemika-dzouchi-no-kami at Kashima and Futsu-nushi-no-kami at Katori; the cult centred on the yatagarasu[7] in the equally ancient Kumano group of shrines; and a cult that may not have survived to the present day associated with the Idzumo line of chieftains known only in the singular personification as Ō-kuni-nushi. All these ostensibly Shintō deities were or are closely connected with the very ancient war deities whose totem was either a three-legged crow or the antlered deer, the latter representing the totem of the Nakatomi clan and later their metamorphosis to the Fujiwara and the Imperial line. Not only that but the yatagarasu totem first appears in the ancient chronicles associated with the war-expedition of Jingō-kōgō when she made a foray to attack her Puyŏ-Kayan cousins in Korea sometime possibly in the fourth century CE. To pursue these early connections further would lead towards a wholly different area than just the roots of military philosophy as it developed in Japan, but it is hoped that it will stimulate further study.

■ CHAPTER 10

The Esoteric Principles Contained in *In* and *Yō*

It is worth a short discussion here to explain the presence of deity figures that are sometimes found near the beginning of some bugei tora-no-maki and denshō, and to point to their connection with the principles underlying heihō. Most experienced warrior students would already be familiar with the symbolism but these ko-ryū founders wished to reinforce the impact of the mikkyō on their students' or successors' minds by a direct visual link in these transmission scrolls.

The concept of 'shade', or *kage*, had in itself a dual meaning and even if not further stated was directly linked to Marishi-ten, the female war deity. In the Yagyū Shinkage-ryū, for example, Yagyū Munenori explains that the 'martial arts' of Japan can be traced back to the husband and wife deities, Izanagi and Izanami. The former is depicted in the *Heihō Kadenshō* flourishing two willow fronds whilst his wife, Izanami, waves a leafy bamboo branch. This is a clear attempt to reinforce the myth that the Imperial line sprang from Amaterasu who was born from the left eye of Izanami. It is also an echo, often repeated, that derives from the sun-deity beliefs of the ruling tribes of the Puyŏ when they began the process of imposing their power in the archipelago.

However, those bugeisha who 'followed the yamabushi paths' would also have been well aware of an additional symbolism contained in the representation of these 'creation deities'. Izanagi figures are sometimes associated with the sun's disc and Izanami with the crescent moon. Occasionally, the disc and the crescent are combined. These are the ancient symbols of Marishi-ten, the 'Queen of the Sky', the female deity of the light which supports sun and moon.[1] The sun figure represents brightness and light, while the female 'crescent moon' figure is 'darkness and shadow'. This can also mean the promise of light to come, the darkness immediately before the dawn; Spring that follows on Winter. If we look at the etymology of the name of one of the foremost Shintō warrior deities, Futsunushi-no-kami, who is enshrined principally in the Katori-jingu, we find that the derivation of 'Fu' in his name is thought to originate from a word meaning 'sparkles of light' deriving from Inner Asia and reaching Japan in the earliest land-taking mythically attributed to the generals Takemika-dzouchi-no-kami (Kashima-jingu) and Futsunushi-no-kami. 'Sparkles of Light' and the 'coming of dawn' are two characteristics closely associated with Marishi-ten. It is significant, therefore, that it is precisely these two warrior-deities that preside over every traditional classical Bujutsu and Budō dōjō and receive the opening and closing *rei* from all present. These two *rei*, by implication, consciously acknowledged or unconsciously performed, pay reverence to Marishi-ten, however subverted the practice and its explanation may be in the present day.[2]

Another interpretation, also advanced within the classical bugei, is that the disc is not just the sun but represents the full moon, signifying *yō*. The full moon is a beautiful and bright object but it is only part of a transient phenomenon, becoming bright but at the same time will lose part of itself. When we see the sliver of the new moon, signifying *in*, we are in darkness that is about to fade as the moon again begins its monthly cycle.

Early morning practice, *kangeiko*, a custom dating far back into antiquity according to many masters, follows the beliefs in the female war deity of moving understanding from darkness towards the light. This concept is directly related to the principles of *in* and *yō*, the use of the unorthodox followed, inevitably, by the orthodox. Sun Tzu stated it thus: 'The resources of those skilled in the use of extraordinary forces are as infinite as the heavens and earth . . . for they end and recommence, cyclical, as are the movements of the sun and the moon . . .'[3]

It would be impossible to ignore the symbolic influence exerted by esoteric Buddhism from its first introduction at the beginning of the ninth century. While it is generally argued that the bushi-samurai class found its origins, according to one authority, as late as the eleventh century, one can only comment that while this group, *per se*, may not have been prominent in its later recognizable form before that date, one can have no doubt that its characteristics were forming centuries earlier, even from the mid-Kofun period in the fourth and fifth century. It was the introduction of such 'deities' as the 'Five Kings of Light'; Fudō-myō-ō, Aizen-myō-ō, Daitoku-myō-ō, Gundari-myō-ō, and Gozanze-myō-ō, and images of the originally Tibetan Maitreya (Miroku), reinforced by Kongaku Busatsu and Zaō-gongen, all Buddhist or near-Buddhist deities with a fierce aspect, plus the violent posture realization of the Ni-ō temple guardians, that must have deeply impressed the warriors and recommended to them the promise that should they become devotees they would enjoy protection from their enemies. Marishi-ten, coming on the scene with her many spells and rituals specifically offering ascendancy over enemies at once accorded with the powers of the foremost of the five 'Kings of Light', Fudō-myō-ō, and was seen, certainly by the warrior elements within the proto-yamabushi, as either a manifestation of Fudō or that he was a manifestation of her. The many images, some of the former possibly pre-dating the Nara period, showing either deity riding on the back of a charging wild boar and associated with the sun and the moon find echoes far away across the Euro-Asian steppe as far as ancient Scandinavia and even Ireland thus reinforcing the pragmatic view that the origin may be sought somewhere in Central Asia.

The Griffin Totem and Marishi-ten

One of the lasting legacies of the Puyŏ migrations that eventually followed one of their Korean high-chieftains' cadet branches as rulers of the archipelago, was the importation of the griffin totem. This seems to have originated deep in Inner Asia in or around the northern Gobi desert and the Altai Mountains between Mongolia and into southern Siberia. The griffin symbol eventually reached Japan with the Puyŏ people, part of the eastern Hsiung-nu nomadic tribes, when they first crossed the Tsushima Straits, beginning sporodically in the first century CE, and sought to establish their hegemony. Their initial settlement attempts were possibly in southwestern Kyushu, but gradually they moved east along the sheltered waterway of the Inland Sea until eventually, and not without having to overcome resistance from the Yayoi already settled in the central region, they established themselves in the fertile Yamato plain, present-day Nara prefecture and north towards Kyoto. A second branch, crossing from the kingdom of Silla in southeast Korea, reached the Idzumo region and effected their rule seemingly with a negotiated settlement and without major confrontation with the local Yayoi tribes.

The griffin-totem warrior group, which may have been numerically quite small, appears to have accompanied the powerful Puyŏ-Kayan chieftains who first landed in Kyushu then moved east. The deer-totem tribe, together with elements of the griffin-totem shaman-warriors, may have first taken control of the Idzumo region, endorsing the rule of the Ō-kuni-nushi ('Rulers of Wide Lands') line of chieftains, possibly seized Kibi in present-day Okayama prefecture, then extended their control through central Honshū to the eastern Kantō. Their final bridgehead was on the Pacific coast of Hitachi province, now Ibaragi prefecture, and the whole of the Kashima-hantō. It was from this eastern group that the Nakatomi chieftains probably emerged.[4]

Returning to the griffin-totem tribes, it seems possible that while small in numbers they had a strong sense of identity. Their rôle in serving the Puyŏ may have been rather specialized providing a religio-shamanic link between the middle world of man and the spirit world of the deities. If this proposal is valid then these men were specialist warriors quite separate from the normal martial tribal groups. In the century or so after their appearance in the archipelago the griffin totem gradually changed to that of the crow or hawk but the original beaked crested griffin image is clearly recognizable on parts of military equipment, particularly many finely-worked intaglio sword hilt pommels.[5] In the late-Kofun period, around the sixth century, these specialist groups disappear but are not singled out in the chronicles as having been destroyed. It is possible that despite their specialized military functions they were not seen as a threat to the now strengthening central Yamato rulers. Perhaps their acknowledged ability to communicate with the deities protected them but, whatever the reason, this shamanic tribe, probably divided into a number of small units, simply took themselves into the mountain regions, and cut themselves off from the lowlands altogether. It may have been at this stage that the griffin totem gradually changed to that of the crow or hawk and here one suspects some additional connection or fusion with the yatagarasu mysteries.

As these reclusive shamanic groups developed within the mountain fastnesses so they became identified with or were the proto-yamabushi who appeared more fully by the eighth century. It is interesting that those later yamabushi groups centred on the Kumano-Yoshino massif and the area dominated by Daisen and the eastern part of Idzumo have connections with the three-legged yatagarasu, whilst the other yamabushi groups, notably in western Honshū, around Hide-hiko-yama in northeastern Kyūshū, and far to the north in the Dewa-sanzan area, are more crow or hawk aligned. By the Nara period, these warrior-ascetic groups were thought of in their former rôle as 'protectors' or 'intercessors' who had the ability to call on the deities of *sankaku-shinkō* at will but now with strong associations with the Taoist 'magician recluses' who increasingly sought metaphysical powers in these remote places. Whatever development took place later towards more regularized Shugendō and an association with the Buddhist mikkyō, an inner and more secretive stream of these yamabushi and their ancient skills concerned themselves with the arts of war that soon focused on the Central Asian female deity, Mārīcī. The development of this deity's cult during the early Heian period brought these particular yamabushi and their beliefs to link firmly with the emerging warrior groups, particularly the *bushi* in the upper level, a connection which became closer and closer as the bugei developed.

This was not an innovation that appeared in the early-Muromachi period but a convergence of two much earlier streams that in some ways may never have parted. The connection is not at a low and broader level but always in the middle or upper levels of heihō, an area of little concern to the emerging samurai class in general but of great significance to those whose birthright was to lead on the field of battle. If, as one leading modern master has affirmed,[6] *all* bujutsu systems sprang from or were totally influenced by the cult of Marishi-ten, then this cannot be dismissed out-of-hand. When we come to examine the principles of warfare that owe their origin to Marishi-ten through transmission by way of the tengu, we find at once that they were based wholly on the teachings of Sun Tzu. This argues that these special shaman-warrior tribal groups under the griffin or crow-hawk totem, were already well-versed in Sun Tzu's *Ping Fa* when they reached the Korean coast and crossed to the archipelago. The exact year that they did this is open to debate, naturally, but suffice it to say it was certainly before the earliest historical appearance of the known Mārīcī texts.[7]

The late Kuroki Toshihiro considered that one of the roots of the major bujutsu, the use of the spear in particular, can be traced back to the dance forms employed in the folk *matsuri*, or festivals. There are many of these stave dances surviving and their purpose is to attract the interest of the deities and 'entertain' them.[8] Nearly all of them make use of the drum, an instrument known in Mongolia and the Altai region as 'the horse of the shaman'. It is also a belief amongst the Buryat tribes of Inner Asia that the deity, when he arrives, actually resides within the drum until dismissed at the end of the ritual; this belief continues in the ritual use of the drum within classical bujutsu and budō. If Professor Kuroki's opinion is correct then the later Sō-jutsu developed in the Yayoi period becoming more sophisticated with the introduction of iron working and straight spears by the Puyŏ-Kayan influx from the mainland. His view is supported by the belief that the iron-smiths who accompanied the early Puyŏ-Kayan immigrants enjoyed a quasi-religious status that associated them with certain sacred sites that later became the focal points for deities patronizing horsemanship and weapon-smithing. It has also been suggested that the *torii* gateways at the entrance to the sacred areas of Shintō shrines may originally have been the frames where horses were secured to be shod.[9] The development of Sō-jutsu in its original aspect of the use of the staff – Bō-jutsu – also seems to have rested by the Nara period in the hands of the yamabushi. If this is the case, as Kuroki believed, the careful growth characteristic of many things going through the 'Japanization' process suggests that it was earlier in the hands of the proto-yamabushi, those shamanic military specialists we have already discussed.

However, the Yayoi people already made use of the Chinese halberd, the '*ka*', a crow-billed hooked staff weapon that it is thought first appeared in China in the early-Zhao period about 800 BCE.[10] This formidable halberd, which sometimes mounted up to three successive blades on its shaft, was, during the Warring States period, employed by both mounted and foot soldiers. On the face of it, the *ka* might be seen as a Chinese innovation that later reached Korea and finally Japan in the mid- or late-Yayoi period but this may not be the case as similar weapons with only slight variations in form were not unknown through central and western Asia, Scythia and as far west as Ireland dating from the Bronze Age of the first millennium BCE. The fact that the *ka* was used by the Warring States cavalry is also of significance, raising the possibility that the weapon and its tactical use may have originated amongst, or been known to, the nomadic tribes of the Hsung-nu *before* it reached China. This supposition is supported by the late H. Russell Robinson's opinion that much weapon technology actually developed amongst the martial tribes living in the northeast of Iran – the Saka, the Alans and the Tochari amongst others. It was these warlike people who disseminated these early arts both to the east and the west.[11]

THE ESOTERIC PRINCIPLES CONTAINED IN *IN* AND *YŌ*

When the use of the halberd developed in Japan it took on a quite different character compared to the use of the straight-bladed spear and so it has remained to the present day. Within the technical nomenclature of both Naginata-jutsu and Sō-jutsu, as in medieval swordsmanship, we can find terms that specifically ascribe the origin of a form or groups of forms to transmission from Marishi-ten, often including 'tengu' and 'kurasu' in the name. Furthermore, in the upper levels of some classical bujutsu ryū we find a number of different weapons employed within the level: sword v. naginata, naginata v. bō, naginata v. yari, etc. It is of interest to note that Yamamoto Kansuke, when he wrote the *Heihō Okigishō* sometime before his death in 1561, advocated that students of the bugei should first study the use of the bō before going on to the spear and halberd. His thinking may have been coloured by an early association with the yamabushi spear skills.

Most Kendō historians agree that whilst we know that these very old arts existed and were transmitted from the end of the Heian period, we have no certain documented records until just after 1500. Some authorities take the somewhat more conservative view and consider that around 1550 would be a safer date. It is a matter of regret that these comparatively late dates should be the case as earlier records, had they been kept and survived the many wars, would have resolved some of the questions, that is certain, and shown a clearer path back at least to the lifetime of Minamoto-no-Yoshitsune. As it is, we can surmise that some of these developed skills existed from the evidence to be gleaned from the *gunki-emaki*, war scrolls, that were produced early in the Kamakura era but we cannot, nevertheless, know how these 'forms' originated or the nature of the transmission in the heihō, only make informed guesses based on close observation and the surviving forms in the oldest ko-bujutsu. In a previous chapter it has been pointed out that such commentary as can be made are at best glosses and at worst either misleading or valueless.

Comparison between the body postures and movements employed in Chinese spear forms and those illustrated in the *Heihō Okigishō* clearly show that there was a lasting Chinese influence on

Sō-jutsu in Japan.[12] The characteristics are unmistakable to the experienced eye. It is a fact that such distinctive movements take a very long period of time to establish and to transmit intact, suggesting that the yamabushi proponents of Bō-jutsu and Sō-jutsu either took great care that change did not take place, difficult where groups were widely separated, or that they maintained some form of interrupted contact with Chinese masters. Suffice it to say that some, at least, of the ko-ryū Sō-jutsu still contains these characteristics despite its complete submergence into the 'regular' warrior bugei which probably occurred in the late-Muromachi.

Naginata-jutsu, on the other hand, may have had its origins in the use of the *ge* as far back as the late-Yayoi but it must be remembered that these hooked-halberds were very different from the *te-boko* that survive in the Shōsō-in repository of the Tōdaiji, Nara. These hand-halberds with their oddly shaped blades certainly seem to be the forerunners of the simple swept-naginata that appeared in the eighth century. The curious weapons in the Shōsō-in were placed there in the seventh century. If the use of the halberd did not originate from the *ge* then it may have been introduced by the Puyŏ-Kayan warriors in the mid-Kofun period of the fifth and sixth century. The spearblades, as opposed to the halberds in this famous repository, do display Chinese characteristics although this may indicate that they were modelled on ceremonial examples reaching Japan from the Asian mainland.[13]

Bugei historians employ the term 'the Four Pillars of Bujutsu' to describe the four main components of the bugei as they emerged in the rather nebulous period of the Nara and Heian periods. The four pillars are: *Kenjutsu*, the Arts of the Sword; *Ba-jutsu*, the Art of Horsemanship; *Naginata-jutsu*, the Art of the Halberd; and *Kyū-jutsu*, the Art of Archery. By the eleventh century when the bushi emerged as a clearly identifiable group, three of these divisions of the warrior skills of the battlefield were adapted both for use on horseback and on foot. However, the skills that were developed by the yamabushi were always those of the infantry, apparently never equestrian, and it should be noted that *all* the skills attributed to the tengu as intermediaries of Marishi-ten were also infantry in nature.

THE ESOTERIC PRINCIPLES CONTAINED IN *IN* AND *YŌ*

From the point of view of tactics and strategy as the understanding of these grew in Japan, it was the use of weapons on foot that best illustrated the principles of warfare. Horsemanship was developed as part of fluid warfare but suitable animals were valuable and needed to be conserved. Their early employment in battle seem to have been mainly by the buke in contrast to the massed cavalry formations that appeared in the later sengoku period.

Perhaps one of the earliest historical examples of heihō illustrated in the *gunki-emaki* is shown in the *Gosannen Kassen Emaki* recounting the Three Year's War of 1086–89 (the War of Eihō) when the Fujiwara clan in the northern province of Mutsu was subdued by Hachiman-tarō Minamoto-no-Yoshi-ie.[14] It depicts with marvellous fluidity a discovered ambush set to entrap the Minamoto warriors as they passed through a landscape of tall shiba grass. The concealed Abe warriors unwittingly disturbed a flock of wildfowl who rise with a clatter of wings thus alerting the Minamoto leader to his danger; in the ensuing chase the ambushing force is destroyed. Yoshi-ie, himself, is reported as saying: 'Some years ago, General Ōe [Tadafusa, 1041–1111] taught me thus: "When soldiers lie in ambush, flying geese break their formation." These have to be enemies lying in ambush.'

■ CHAPTER 11

Unexpected Attacks Against an Unprepared Enemy

During the Gempei War at the end of the twelfth century, the sudden and completely unexpected attack on the retreating Taira army by Minamoto-no-Yoshitsune at Ichi-no-tani is an excellent example of *cheng* force changing to *ch'i*. In this famous incident that occurred in 1194 where the apparent harassing strategy of the Genji in pursuit of the Heike was changed by the brilliant young Yoshitsune to a mounted attack launched down a precipitous declivity, the enemy force were caught in the midst of embarking from a narrow beach in the Bay of Suma, just to the west of present-day Kobe, and were severely mauled. During the attack there took place the fatal encounter between the Genji warrior, Kumagaya Naozane, and the young Heike lad, Taira Atsumori, the subject of a famous Nō play[1]. Even today, just over eight hundred years later, the place where Atsumori's head was displayed to Yoshitsune and the grave of this young bushi still has the power to move strong men to feelings of deep emotional regret. However, this romantic account highlights the lesson contained in the brilliant application of ch'i by the youthful Yoshitsune. The steep slopes descended by the Genji horsemen look formidable even today and the attack must have disconcerted the Taira on the strand below exactly as the accounts describe.

Minamoto-no-Yoshitsune demonstrated his grasp of the principles set out by Sun Tzu by realizing that the Taira embarking from Suma beach were dispersed and in no position to reassemble. He saw at once the chance of a surprise attack delivered to the heart of the enemy if he and some of his men, after a lightning pursuit, could find a way to descend from the sloping heights. He followed the precept in the *Ping Fa* of 'speed is the essence . . . travel by unexpected routes and strike (the enemy) where he has taken no precautions'. It was a case, as one of Sun Tzu's commentators, Chang Yü (late Sung period), observed: 'that the one thing esteemed is divine swiftness'.[2]

This same principle was applied at the famous attack delivered by Oda Nobunaga against a far superior force commanded by Imagawa Yoshimoto on the twenty-second day of the sixth month in 1562. Threatened by imminent attack by the Imagawa army which outnumbered Nobunaga's force by ten to one, Nobunaga decided that his only course of action was to launch a surprise attack. It was reported that Imagawa Yoshimoto had decided to bivouac overnight in a constricted valley named Dengaku-hazama which lay some distance to the east of present-day Nagoya. Certain that his army would not be attacked because of its obvious strength, Imagawa chose this site to rest his men before crossing the border into Owari proper. A violent rainstorm hid the small force of the Oda, reputed to be about five-hundred strong, as they worked their way round to the hilltop at the rear of the valley. They descended the slopes and charged full tilt into the Imagawa camp. In the confusion and panic, Yoshimoto was surrounded and his head taken, later to be presented to Oda Nobunaga.[3]

It is possibly with Ichi-no-tani and Dengaku-hazama in mind that the Japanese military command and Admiral Yamamoto planned the surprise attack on Pearl Harbor in December 1941. The American Pacific Fleet lay in Honolulu Harbor secure in the knowledge of its might and lulled into a false sense of security by the fact that America was not at war with Japan – yet. The Japanese air attack, made at low level to avoid giving clear advance warning, was devastating as is well-known. If it had been followed-up, who can tell how long the Pacific War might have lasted?

In all three of the examples of ch'i surprise attacks above, the larger and more powerful side failed to post adequate watchers who could give advance warning. In addition to being caught unawares, the commanders neglected gathering intelligence and paid the inevitable penalty.

'Threaten in the East; Strike in the West'

In this discussion of the medieval understanding of 'ch'i' and 'cheng', 'in' and 'yō, the writings of Mao Tse-tung serve to illustrate how fundamental and important are these principles in modern warfare. The Japanese bugei of the Muromachi period and the application of Sun Tzu's dictums at that time were well-documented but how these matters and their lessons were transmitted and their lessons assimilated is more difficult to understand by the very nature of these military affairs. The term 'in' as we have seen, implies 'hidden' or 'secret'; these were matters not to be disclosed to all and sundry and certainly not to a potential enemy at any time.

The principles contained within the Japanese martial tradition are not concerned with the precise technology of the weapons used, they are the application of concepts regardless of the size or composition of the fighting units. Much of what Sun Tzu and his fellow philosophers suggested is just as valid today, despite our highly advanced military technology, as it was two-and-a-half thousand years ago. The teachings remain locked inside these entities to be found and understood by those who have the ability to peel away the outer layers to reach the innermost core. The problem of reaching this degree of understanding is that the method, then as now, requires dedication, determination, time and sincerity. It was the latter quality that was always most prized by masters of the bugei. Few men are as gifted in this perception as the youthful Minamoto-no-Yoshitsune; most have to undertake an exacting and gruelling apprenticeship first training the body and then the mind.

The principles we are examining here are those that are contained in the oldest of the bugei insofar as we are able to maintain the argument for their integrity. It is important that we do not permit any taint deriving from the modern tendency to equate the bujutsu with budō, and therefore with sport, to influence our evaluation. While there is nothing wrong with the popularization of some of the later developed entities in this direction, here we are trying to strip away any modern connotation and return to the original sources within Japanese warfare itself and particularly to see how the expert bugeisha understood heihō.

How was the principle used for this chapter's title explained in ko-bujutsu? In a number of ko-ryū this is illustrated by the composition of each 'level' making up the whole. The basic principles are first introduced in the 'ōmote-dō' kata and explained, often in terms that exercise the deshi more physically than intellectually, then subsequently the later forms vary and add to these basic concepts in order to broaden the student's mind and stimulate his awareness. We need to remember that most students who sought out an expert master in the Muromachi period probably did so after gaining a high degree of basic practical experience through their own experience of warfare. Hence the criticism made by Ōmori Rokuzaemon Masatora, the late-seventeenth century master of the Hasegawa Eishin-ryū, that the young samurai who came to practise Iai in his dōjō were nowhere near as experienced in these arts as their grandfathers had been.[4]

Simply stated, and keeping in mind that the following comes from the field of Kenjutsu-no-heihō, we can illustrate 'disturbance in the east; attack in the west' by the following sequence of cuts used in one tradition's 'summing-up' form that summarizes the first series of kata moves. At the outset, both swordsmen make a feint jabbing lunge forwards, an opening termed *'jibari'*, a move that to some may appear to be just a formality but in fact represents the initial opening probe by field commanders making contact with the enemy. Shidachi at once makes a thrust towards his opponent's chest which elicits Uchidachi to draw back, sword coming to his right shoulder, *in-no-kamae*, and replies with a very fast cut at Shidachi's head which is avoided with a low left step and instantly Shidachi whirls his sword up past his left shoulder and cuts *shomen*. Missing, Shidachi turns away to his right, feet together, knees flexed down, with his sword across his lower belly in a posture known as *hara-tombo*. Uchidachi cuts head again but this cut is met across Shidachi's blade, his left hand supporting the blade, and the attacking sword wound violently down to Shidachi's right. A sharp cut to Uchidachi's right wrist is blocked and Uchidachi pins down Shidachi's blade horizontally. Unexpectedly, Shidachi sharply throws up his blade and without any warning makes a horizontal slice low across his opponent's groin with the intention of severing one or both of Uchidachi's femoral arteries. If he misses he circles his sword round and up past his left shoulder and makes a lightning cut at Uchidachi's right neck . . .

After the initial *jibari* all this action takes place at a bewildering speed, the performance including three circling loops first to the left, then to the right, and finally back to the left, constantly forcing the opponent onto the defensive and never permitting him to settle. The form continues with similar fluid movements accompanied by changes of tempo that contain, albeit in modified form, moves already practised earlier but here presented quite differently in order to stimulate the intuitive understanding of the deshi.

While this is admittedly a complex sequence of feints, cuts, changes of direction and a final attack, it is emphasized that it is only the *beginning* of a much more involved form relatively early in the first level of this very old and mature ryū. On analysis, the cunning structure of these early moves reveals a series of gambits that are intended to confuse and surprise the opponent. The purpose of the *jibari* by Shidachi is to cause the enemy to break off his own preparations and regroup in order to deal with the threat and make a head cut. Uchidachi does this on the principle of dealing with a low attack by countering with a high one of his own. Unfortunately for him Shidachi has moved left, circled his sword to confuse, and cuts very fast at Uchidachi's right head (or neck) then moves with a large step to the right, turning sideways to his opponent, sword held *hara-tombo* at his right hip. This is to invite a second head cut from Uchidachi . . . It is a feature of this particular tradition to use hara-tombo prior to delivering another attack; in the case in point the intention is to lay a trap, to lure the enemy to deliver his cut just when and where Shidachi wishes it to be.

Two important principles outlined by Sun Tzu are demonstrated at this juncture: the first is to 'Appear at places to which the enemy must hasten; move swiftly where he does not expect you'; and the second is the fundamental one that 'War is based on deception. Move when it is advantageous and create changes in the situation by dispersal and concentration of forces.'[5] Shidachi's jibari is similar to the arrow shot often used prior to joining battle and the custom might be traced back to Yayoi times when a sacred deer might be selected for sacrifice by the deity 'guiding' the flight of the random arrow discharged towards the herd by the shaman-priest. The similar discharge of a random 'spirit-guided' arrow has been noted amongst the Buryat nomadic tribes of Inner Asia, descendants of the Hsung-nu horsemen. The hara-tombo trap is sprung when Uchidachi again cuts at Shidachi's head; the cut is met with a sweep upwards to cut the attacker's descending wrist and the two protagonists end up with sword blades locked out to Uchidachi's left, his blade pressing down horizontally on Shidachi's, both men shoulder to shoulder. After a momentary pause, Shidachi creates further 'uproar in the east' by jerking up Uchidachi's blade and slicing horizontally across his opponent's groin threatening severe injury or worse . . .

To summarize, the swordsman has, feinted to the front then with a series of fluid moves both to his left and right and back again, totally confused his opponent and caused him 'to defend places he cannot defend'. In terms of the 'four poisons' of heihō, we have 'confusion', 'surprise', 'fear' and 'doubt' all present and in that order, and all contained in a form that dates from, at the very least, the first half of the sixteenth century but probably considerably earlier.

Cheng to Ch'i

The series of cuts and feints described above generally demonstrates changes in direction from low to high and side to side; the following example taken from Iai-jutsu exemplifies directional changes from high to low. This, too, is from a very early tradition of Battō considered to date from the early part of the Sengoku period.

The swordsman has approached his enemy and comes to a halt facing him; suddenly, and without warning, he draws out his sword low and cuts across at about waist height to his right, whirls the swordblade up past his right shoulder and strikes across to the left at about thigh height, double-handed. At the same time he drops down, left knee kneeling. Without the slightest pause he cuts back again horizontally at about the same height and, coming to his feet whirls the blade round his right shoulder, shakes off any clinging blood with a strong movement known as *kiri-te* to his right side, and returns his blade to its *saya*.

Here, instead of moving to the right or left, the ch'i is the sudden change of direction from the low belly cut to high and then low from right to left, dropping to left kneeling. While the main principles involved are circling and changes of direction, provided the master judges his timing correctly then his opponent is mentally off balance and finds it impossible to counter-attack effectively because to do so he must first defend everywhere at once. Again, judging the moment correctly, the ch'i attack is unexpected and devastating.

A somewhat grim axiom has come down to us from the classical bugei: 'A good swordsman requires two cuts to dispose of his enemy; a true master needs only one!'

Ch'i to Ch'i

The interpretation of the 'four poisons', as they are described, are many and varied and as might be expected there are countless examples to be found within the ko-ryū; once understanding has been achieved then the heihō-sha can turn to the application. The combination of surprise and doubt is a powerful one but, like our second form, when vertical movement is added, it becomes truly formidable. While familiar with two such forms, just one will be sufficient to illustrate the principle.

The swordsman positions himself before his opponent, expressing not the slightest threat in his posture. Quite suddenly, sensing a change in his enemy's mind, he loosens his *tantō*, or dirk, in its sheath and with his left hand suddenly draws it out and tosses it high into the air above his opponent's head. Involuntarily, the enemy draws back, understandably fearful of the razor-sharp blade, but the swordsman is already cutting into his chest or, if both are in armour, across his exposed throat. There is no escape. It is, in very simple terms, an excellent demonstration of 'moving east to strike west', of disorienting and unbalancing an opponent to create an instant when the strike will be the most effective. In Iai-jutsu, as we have already noted, such demonstrations of principle are always brief and never continued to complexity except in rare situations where the opponents are many and there is no other solution other than to act.

'Ch'i' movements, then, are brief and fleeting which is the reason why heihō explained them in teaching as like 'curling smoke; visible yet intangible'. Even the parabolic curve of the tossed tantō, brief as its flight was, had no real form or shape. The surprise and shock implicit in such an unexpected threat demanded full attention yet, in truth, the greatest immediate danger was posed by the swordsman's long sword. No further moves were even considered by the master who perfected this tactic, only attention to *zanshin* in withdrawing from the situation. What might follow, in the opinion of many masters of Iai-jutsu would fall into the realm of Kenjutsu. At this point ch'i changes back to cheng.

Diversionary Tactics

Creating the chance for an attack would usually be considered to be more in the field of conventional warfare but some circumstances require the unorthodox. In Iai-jutsu this is sometimes addressed with probing infiltration in mind and forms include simulating night patrols sent out into enemy territory to obtain information or, of course, infiltration designed to create the chance of further guerrilla activity. One such form involves dealing with an enemy patrol encountered during just such an intelligence gathering foray. The swordsman, his senses acutely alert with the foreknowledge that he is on enemy ground, silently creeps forward and becomes aware of an opponent just ahead of him. He takes a large turning step to his left, his left knee just brushing the ground and body very low, drawing out his sword to the right and sharply, at arms length, taps the *kissaki* on the ground at full stretch to his right. At this sudden unexpected noise his startled enemy cuts at the sound in the darkness. The swordsman, a full pace to his opponent's right, turns in and cuts about a metre to the left of where he struck the ground, injuring or severing his adversary's arms or even his torso, it hardly matters which. The diversionary tap on the ground may be made directly to the swordsman's right or to his right rear but within this ryū there are at least two variations.

The stooping almost to half-kneeling is not only to avoid being seen by the approaching enemy but to lower the kenshi's horizon sightline so that the upper part of his opponent's bulk may be seen against the night sky. It is of great importance for the swordsman to be able to judge precisely when to strike. In this ch'i form emphasis is placed on the foot movement and achieving a silent approach.

The same comment might be made about another 'infiltration' form, one that this time uses the descriptive metaphor of a tiger stalking its prey. The difference is that this form examines the problem of dealing with an enemy and then, knowing that some sort of alarm has been raised, of extricating oneself from what could prove to be a difficult and dangerous situation. The swordsman advances silently into enemy territory, i.e. he moves across the ground with pointing noise-absorbing steps based on close observation of how a tiger moves with the pads of his paws soundlessly masking any sound. He realizes that somewhere close ahead of him is an enemy, recognizes his position and suddenly draws his sword to cut the man down; the second cut is made kneeling to disguise the action from any alert observer. Now is the time to withdraw but the kenshi takes care that when he rises, still keeping low, he takes his first step back with his left (rear) foot in order to feel that the ground is free from obstructions. In the form, the swordsman takes six backward steps, still exercising great care in how he does so. At this point, knowing that he is pursued and still in a flexed lower-than-normal height posture, he recognizes the nearness of his pursuer. The kata ends with either a sudden drop to one knee whilst cutting down the second enemy, or with an equally unexpected vertical leap whilst drawing sword on the second man, a leap form luring the enemy into an ambush. This leap is given the name of *'tengu-tobi'* and truly constitutes a surprise ch'i tactic. It is based on making full use of intelligence already gathered during the night foray into enemy territory.

Striking from Concealment

From the Kenjutsu of the Muromachi period , and again possibly much older, comes the ch'i tactic of ambushing an unwary enemy then bursting out on him with a brief and devastating attack. 'He who is prudent and lies in wait for an enemy who is not, will be victorious.'[6]

In former times many highways and tracks passing through flat marshy areas like the Kantō plain in eastern Japan were bordered by vast swathes of tall *shiba* grass which often grew above the height of a man. In this form, the kenshi, seeing his enemy approaching, steps into the tall grass, sword drawn, and with a large circling motion gathers in a sheath of the grass in front of him, thereby effectively masking his presence. His swordblade is held vertically, edge towards him, just before his face, wrists rotated inwards. The instant his enemy is within striking distance, the kenshi 'uncoils' and thrusts forwards out from his cover, point directed at his opponent's throat. If the strike is fractionally premature, he thrusts again . . . In such a situation there can be but one result!

* * *

Concealment can imply the setting of an ambuscade, of course, but in a more general way the chief preliminary characteristic of Iai-jutsu, and the one that continues into the later development towards the Iai-dō systems, is the non-aggressive posture and demeanour of the swordsman prior to his draw-strike. The half-sitting posture, *tate-hiza*, shown by advanced masters suggests that they have almost fallen into a state of quiet contemplation, they are so calm and relaxed. Hayes employs the term *'Mukamae-no-ri'*, 'the inward understanding of posture', to describe one of the 'Eight related characteristics of the fighting arts' and the late Okada Morihiro *Hanshi* explained it as *'mu-no-kamae'*, having 'no-form', neither showing weakness nor strength. It is a state of being able to evaluate a situation and deciding on action or no action before it is too late.[7] While Iai-jutsu was devised to exercise the swordsman in the principles of unorthodox action in heihō, the underlying truth follows Sun Tzu who taught that 'those skilled in war subdue the enemy's army without battle'.[8] This is expressed in the Japanese bugei as *nuki-ai*, to achieve victory without fighting and is at the root of the important concept of *katsujin-ken*, the sword used to preserve life.

The half-kneeling *tate-hiza* posture is cultivated within all the ryū deriving from the inspiration of Hayashizaki Jinsuke and is demonstrated at its simplest in the form entitled *Yokogumon*. This name has two levels of meaning. The first and most commonly explained, suggests that the track taken by the sword is like that displayed by the flat-bottomed clouds seen on a pleasant summer's day. Many such small cumulus clouds have a flat base and a narrow lozenge-shaped body above as they hang lazily in the summer sky. In *Yokogumon*, this is the image in the swordsman's mind when he draws on his opponent. There are no histrionics, just a smooth deadly horizontal strike half-stepping forward from tate-hiza or a standing position. The draw-cut is followed by the blade being whirled back past the left shoulder and ear, and a slightly flattened final cut to despatch the enemy. It is held important that the 'flat cloud' image is always visualized. This form may seem quite simple on the surface but, as in many traditions within ko-bujutsu, it sets out only the basic requirements of much that is to follow. As such, it is the foundation form.

The second level of meaning may have become attached to a similar movement with the sword long before Hayashizaki Jinsuke rationalized the structure. Here, we can see a connection with Taoist magical incantation rituals designed to add metaphysical power. Just how old this connection may be is far from clear since the theory may have developed from the secret ritual sword-incantation 'dances' of the yamabushi or ancient Shintō. At the beginning of the next level, the first of two series that contain the inner teaching principles of heihō, the name given to a similar form is *'Kasumi'* which means 'mist', 'hazy' or 'obscured'.

Essentially, the preliminaries and draw-cut of *Kasumi* are identical with *Yokogumon* but the following technique then 'creates confusion in the east and strikes to the west' before a direct finishing *coup-de-grâce* cut to the front. Again, this is 'ch'i' followed by 'ch'i'. The imagery behind this form is extremely interesting since it is quite clearly based on an incident that involved the semi-mythical third son of the Puyŏ-Kayan chieftain, Keikō, probably in the third century CE. The prince was once ambushed and surrounded by hostile tribesmen whilst sheltering for the night in a small hut. The 'barbarians' set fire to the tall *shiba* grass with the intention of forcing him out or roasting him alive but Yamato-takeru drew out his 'magical' sword and with strong scything swings cut the grass down in wide swathes about him, the stems lying in such a way as to guide the hungry flames away from the centre. This blade, aptly named 'Grasscutter' is said to remain one of the Imperial Treasures. While the legends surrounding Yamato-takeru closely echo those of Bairam in Iran and our own Arthurian legends in the West, the imagery contained in *Kasumi* is quite extraordinary.

Strike Where the Enemy Does Not Expect

The master swordsman of the Sengoku period clearly subscribed to Sun Tzu's simple dictum: 'Come like the wind, strike like lightning', seeing matters exactly as Mao Tse-tung did in recent times. The kenshi directed considerable energy into devising what, on the surface, would appear to be relatively straightforward tactics to deal with two or more opponents, often positioned at oblique angles to the swordsman. In analysing these moves we must remember that, in military terms, the simpler the principle is demonstrated the better it can be absorbed and applied. The complex is always more difficult to understand and is prone to defects where unexpected factors may come in.

To give an example from the early-sixteenth century, the swordsman, walking along and minding his own business, becomes aware of either pursuit or an imminent threat from the rear. He takes a quickened step forward, thus substantially increasing the interval between him and his assailant, at the same time drawing out his longsword. He crouches whilst turning left to face the rear, and thrusts forward and upward towards his enemy's chest or throat. The thrust is low to high. Grasping the *tsuka* with both hands, he whirls the sword round to left *jōdan* and strongly cuts down, *kesa-giri*, from his left to right. The action, once it commences, is fast and devastating.

Another Iai-jutsu that employs the same principle of ch'i deals with a similar threatening situation, this time rather more immediate. The kenshi, realizing that his pursuer is about to deliver a head cut from the rear, turns to his left drawing out his blade clear of the saya and leaning forward at the same time in order to confuse his enemy's judgement. He whirls completely to his rear and, grasping the tsuka with both hands, cuts back and upward with reversed kesa. Continuing the arc of the sword over his left shoulder and turning his hands, he cuts back down, *kesa-giri*, to his opponent's right shoulder. These two cuts would certainly sever the enemy in two. The second cut, on the other hand, may well be intended to deal with a second assailant.

Once again, in this early Iai-jutsu form we can see the use of misleading, circling steps and the unexpected oblique upward angle of the first cut. The force with which the upward *kesa-giri* is delivered would clear the newly slain opponent out of the swordsman's path to deal with any further attacker.

* * *

Another form, also deriving from the eastern Kantō, that dynamic centre in the development of the bugei during the mid- to late-Muromachi era, makes use of fast stepping action coupled with a *tengu-tobi* jumping turn followed by a thrust delivered from a deep crouching posture. The underlying thinking behind this form is intended to illustrate Sun Tzu's dictum that 'when an enemy pretends to flee, do not pursue'.[9]

The swordsman, in retreating away from his opponent, is able to conceal his hands as he begins to draw out his sword. Sensing the proper moment to strike has arrived, he completes his draw whilst making a high tengu-tobi through 180°, turning to his right. At the same time he cuts down obliquely, left to right, ending in a low crouching position, sword held hara-tombo at his right hip. This form, like at least four in the Hasegawa-ryū corpus, is one of the clearest examples of a 'coiling smoke' ch'i technique. The movement whilst confusing the opponent adds impetus to the downwards sword stroke.

Hayashizaki Jinsuke, later in the Sengoku period, devised two not dissimilar forms within his own Battō-jutsu, the first of which he named as 'Scale-stripping' and the second 'Rock amidst breaking waves'; neither term giving much away to the uninitiated. In these forms it is clear that this master swordsman was applying the principles of not being 'shaped' with that of Marishi-ten's coiling 'invisibility'. In 'Scale-stripping', the swordsman is half-kneeling, facing right, and is attacked by an enemy approaching from his left. He rises and, turning, cuts his opponent partly obliquely across his chest, whirls his sword up left to jōdan and, dropping again to a very deep posture, cuts the adversary from left shoulder to right hip or simply strikes just above his left hip. It is the swordsman's circling motion once he begins that creates confusion since the enemy, in his moment of recognizing that his intended victim is moving, is unwittingly deflected by the tiniest amount to his own right front. At the same time, the kenshi is rotating anticlockwise which takes his centre line slightly right and his whole left side, the side still at risk, away from left to right. The kenshi's final swordstroke is delivered fully *hanmi*, or half-turned.

The second of this pair of forms is often described as an 'alternative' one but this is not the thinking of a number of masters although it is true to say that several variations do exist within this group of Iai ryū-ha.

The swordsman senses an attack threat from his rear as he sits tate-hiza. He rises and draws his sword as he uncoils, again rotating to his left, and cuts up under the attacker's right arm, whirls his sword left to jōdan, pauses momentarily then, dropping to the same low crouch as in the last form, he cuts his enemy just above his left hip.

The path followed by the sword in the illustration of this form shows the 'coiling smoke' trajectory, one which misleads the opponent's judgement causing confusion and loss of purpose and creating the chance for a devastating counter-attack. These two forms must have afforded much satisfaction down the centuries as they show a natural beauty in their movement and are, to many masters, the perfect Iai. The fact that they have each acquired so many variations attests their significance in the minds of some kenshi. In terms of heihō these are 'cheng' to 'ch'i'.

Coiling-Smoke Principle Applied

The master-swordsman, Hayashizaki Jinsuke, reputedly based a significant proportion of his forms on his original intuitive inspiration of a rearwards thrust. Either he or one of his immediate successors, possibly Hasegawa Eishin, according to one tradition within the ryū, devised a highly original tactic that utilized the 'coiling' principle and placed it firmly in the hands of a cunning general.

In simple terms, the gateway entrance through a defensive work should, in theory, be well guarded and furnished with clever devices. Unless an attacker has great strength and launches a direct attack in clear view of the defenders, the guards placed to control such a point will be able to hold the position until the alarm is raised and assistance arrives. Hence the placing of the guardhouse just within this strongpoint in many military forts, small or large, in all parts of the world, not only Japan. However, if the attacker is very much weaker than the garrison, he will not even consider an assault on so strongly held a place but will seek to penetrate the perimeter elsewhere. All this may seem self-evident but the main gate entrance is often only assumed to be strong; the reality can be quite different.[10]

The ratio of guards to the potential attacker in this particular example of Iai-jutsu is no less than 3:1; quite significant odds, one feels.

The single swordsman, seemingly innocent, walks up to the gateway where there are three guards deployed in the normal approved manner; that is, one slightly in advance of the gate and two guarding the actual entrance itself. He passes the first sentry to engage the attention of the other two, all in a deceptively normal manner. At this moment, the kenshi starts to turn away as though retracing his steps, draws low as he turns but reverses this movement and thrusts the nearest of the gate guards, whirls and cuts down the outer sentry behind him, whirls again and cuts down the remaining guard leaving him in complete control of the very strongpoint that the enemy should have protected at all costs.

This ch'i tactic breaches the defence at the point least expected. An excellent example of this tactic in action was the sudden and totally unexpected surprise when the Israeli armour recaptured the Golan Heights in the 1967 Arab-Israeli War. The road climbing to the Heights was heavily fortified and was the last place that any attack was thought to have any chance of success. The sudden thrust took the strategic Heights with ease.

The first principle of 'Through a Gateway' was the clearly 'normal' approach where the swordsman could not possibly pose a threat. He appeared weak and his appearance was therefore deceptive. He was indeed weak *but* he was still armed. He passed the outer sentry with no trouble because of his confident everyday demeanour that sent neither positive nor negative messages to the guards. He was able to reach striking distance without a hint of what was to come. In turning away as he starts his attack, the circling movement cannot at first be shaped by any of the three enemies but the danger is the unknown quantity of the outer guard's reaction. The order in which the theoretical form deals with the enemy is formal but is based on logic. He thrusts the nearest of the two guards, whirls left and cuts down the man to his rear, whirls back again and cuts down the third guard before he can regain his composure from the shock of seeing his comrades killed or call for aid. Finally, the swordsman has no need to retreat; his 'ch'i' has become 'cheng' and he awaits his reinforcements.[11]

Divide and Rule

Reducing the odds is an important matter in warfare but a necessary one since by dividing the opposing force the commander is faced with two radically weaker units. If a commander faces a united enemy numerically twice his own force's size and strength then he might expect at best a heavy mauling; at worst, a defeat. But if the enemy can be divided into two separate parts by deception, victory might be snatched.

Hayashizaki Jinsuke placed two seemingly 'cheng' forms at the beginning of the upper level of the *kuden*. These ostensibly examined the problem of approaching two enemies in a constricted space. The first form, from the heihō viewpoint, looks at two enemy forces converging and a tactic of dealing with them before they can effect this juncture. The second form takes this problem further and is a suggested tactic against the possibility of the two enemy groups attempting to encircle the smaller 'defending' force. In both forms, remembering that the situations are greatly simplified to their lowest common denominator, the kenshi is in the open and can be 'shaped' but must act decisively before the enemy realize their danger.

In the first situation, therefore, the swordsman approaches his enemies apparently unconcernedly. He must get close enough to be able to strike. He walks forward as though to pass between his opponents. The configuration can be described as the letter 'Y' with the swordsman walking up the stem and his enemies approaching on the branches. When he reaches cutting range and before the other parties have joined, the swordsman suddenly draws and cuts obliquely to his right-front, severely mauling or destroying the opponent on that side; then he whirls up his sword past his right shoulder and, momentarily gathering himself, he steps in left-front and cuts down the second man. To achieve 'formlessness' as he cuts the first man, he slightly sways in the direction of his one-handed cut thereby making it more difficult for the second opponent, if he is alert, to judge exactly where the kenshi is positioned. The second cut, double-handed, is the strongest of the two, of course, on the principle that if the first man is not rendered *hors-de-combat* then all is lost anyway.

In the second of these two forms, the kenshi moves forward until his opponents are on either side of him. It is as though he has moved up the stem of a letter 'T'. When he is within striking range he draws his sword out rapidly towards the opponent on his right, threatening to strike him with his *tsuka-koshira*[12] and swaying appreciably towards him, then he thrusts the man on the left, whirls his sword up past his left shoulder and ear, pauses fractionally, then cuts down the first man. As in the first form his swaying motion makes it difficult for the enemy to 'shape' him. Contained in this ch'i 'divide and rule' is sound reasoning in that the enemy, split into two weaker forces, must exercise caution lest either cuts his ally, whereas the kenshi can strike at will having no such constraints. Later on, within this upper part of the *kuden*, the swordsman examines methods of dealing with opponents to his front and rear, or those obliquely placed, all the forms being relatively simple and, once mastered, highly effective examples of heihō.

Perhaps it should be noted that the usual explanations given to students of the above two forms, and a number of others considered here, in the earlier stages of their training make no mention of the wider tactics involved but are limited to the more mundane presentation that they are for simply dealing with two 'escorting' armed men.

Making Use of Cover Prior to a Surprise Attack

Most Iai forms at first glance appear to contain pre-emptive action. This is the core characteristic of techniques that demand action from a very close space. Some forms, however, identify an enemy from a distance and employ decisive unorthodox steps to deliver a cut, usually only a single one, as soon as the swordsman is within range. Once again, the following examples of these tactical forms are taken from the heihō developed by Hayashizaki Jinsuke in the second half of the sixteenth century, and entail, at this basic simplified level, dealing with an enemy in a crowded street. Anyone who has travelled in the older quarters of Japanese towns or villages will know just how narrow and congested even some main thoroughfares can be.

In the first of these forms, the swordsman glimpses his enemy some way ahead within the throng. So as not to alarm any passers-by, he creates a slight space by stepping back a little drawing out his longsword as he does so and grasping it, point to the rear and edge upwards, at the side of his left shoulder. His arms are folded high across his chest with his elbows almost joined ahead of him and at chin height. Rushing forward impetuously, he bursts through the last of the crowd, arms and sword sweeping all aside, and strikes down his enemy from jōdan.

This is 'ch'i' literally with a vengeance. The sudden impetuous charge violently thrusting aside the last of the intervening crowd is intended to cause confusion and possibly indecision, all that is needed to make this highly unorthodox attack effective.

Following this 'Sleeve-brushing' form we have a second one, designated as an alternative but of quite a different character. This form seems to have lost its original name, although it may never have been given one or the title did not survive the traumatic period that followed Bakumatsu between 1877 and 1900.

The kenshi once again picks out his enemy amongst the crowd in the thronged street and, as before, steps back with a pause, drawing out his katana low and holding it well back against his *saya* at his left hip, *kissaki* to the rear. He rushes forward, not directly at his opponent but a little to the man's left. With a strong slashing cut low and to his right in order to clear a space, he whirls the blade up past his right shoulder, changes direction to his original left front, and steps in cutting down strongly from jōdan. The basic concept behind this drastic but unorthodox form is 'uproar in the east, strike in the west'.

It is of interest that in the majority of such 'east-west' actions that I am familiar with, the tactics are precisely that – feint right, attack left. The argument advanced by way of explanation for this, within my own experience of bujutsu, is that in facing an armed opponent it is prudent to draw the sword for the initial cut with a step outside the opponent's left side, thereby partially reducing the chance of being wounded by his blade should he draw at the same moment. While this seems reasonable when given as instruction to relatively inexperienced deshi, I believe the true reason is that the draw cut must be made with movement that will create momentary difficulties for the enemy to 'shape' the swordsman. 'Move > strike > move >' is fundamental in all close-space combat at any level and in any form.

■ CHAPTER 12

Fundamental Teachings

In certain areas of ko-bujutsu where many, if not all, of the original concepts have been preserved, the observer is somehow aware that the entity he is observing is very different from the formality that overlays so much of the later budō. Here, the movements flow subtly and seamlessly into each other and yet convey a crispness and clarity that only comes from long practice and experience. To use a modern analogy, it is the difference often obvious between amateur and professional boxers even where the amateur is at the height of his powers. Of course, there are notable exceptions, but in general the difference in quality is always to be seen.

This flexibility through the use and application of natural movement is especially clear with, for example, the Iai-jutsu within those ryū-ha grouped around the original inspiration of Hayashizaki Jinsuke. It is also a characteristic of many surviving bujutsu traditions, especially those originating in the eastern Kantō region. Where the integrity of these traditions has been maintained without permitting any later influences to weaken and distort the original, we can still have a valid 'window on the past'. It is in these entities that we find the strongest sense of timelessness.

This quality of natural and closely-observed movement, sometimes expressed by masters as 'at first slowly, then faster, and finally like the wind', surely derives from Sun Tzu's aphorism that 'subtle and insubstantial, the expert leaves no trace'. In these entities we can find true 'art'. The movements in many of the forms, no matter what weapon is employed, contain what the eighteenth-century English artist, William Hogarth, described as 'the line of beauty' in their harmony of composition. Because the movements are natural they become deceptive; their simplicity gives them focus and the ability to blend, dissipate and reform just like coiling plumes of smoke rising into the still early morning air. Their naturalness prevents them having concrete 'form'. When this quality, thoroughly absorbed, is applied to the use of weapons, particularly sword and spear, it becomes deadly. The heihō-sha, the expert strategist who could further project this art to the field of warfare, was truly one to be reckoned with. In these principles, irrespective of technological advances and innovations, understanding art is essential. 'Cheng' is the former, the surface technology and its disposition or organization; 'ch'i' contains the art that cannot be quantified if it contains Hogarth's 'line of beauty'.

The 'orthodox' in some aspects of warfare and the underlying bugei, is the disposition and use of force and expressed by such phrases as *'shishi-funjin'*, the impetuous overwhelming attack of a lion. The 'unorthodox' ch'i tactics are subtle and deceiving; they are the movements revealed to man by means of the tengu often using the metaphor of the powerful tiger – seen by many but rarely understood.

Deception in the art of war means invisibility. The lack of 'shape' does not mean disarray or weakness but, properly applied, provides strength. Again and again Sun Tzu returns to this theme and the importance of this principle is taken up and emphasized by numbers of medieval Japanese kenshi and such modern exponents as Mao Tse-tung, particularly in the area of guerrilla warfare. 'Cheng' tactics are relatively easily understood and absorbed by the competent student of warfare; it is 'ch'i' that requires much deeper insight in order that the unorthodox might be applied. Deception, intelligence and subtle movement; all three combine and blend like coiling wisps of smoke, confusing the enemy and causing doubt and indecision, finally depressing his morale and creating fear. It is no wonder that in medieval times even realists like the military commanders and the kenshi appealed to higher invisible powers in the spirit world such as Marishi-ten and Fudō-Myō-ō for their guidance and added protection, or believed that they could receive their inspiration through the medium of dream-revelations or the tengu-yamabushi. It is the reason that we find, if we are prepared to enquire deeply enough, that belief in these deities and their 'messengers', like the tengu, underlie every ko-ryū of the classical heihō.

CHAPTER 13

Warfare and Ritual

A recurring part of this study of the medieval understanding of tactics and strategy as exemplified by the Japanese application of Chinese military classics is the close links in many warriors' minds between the metaphysical and physical levels. While Sun Tzu clearly advised against putting faith in deities or augury of any description,[1] nonetheless there can be no doubt that such practices were held to be vital at all levels of conflict right from the beginning of the first Puyŏ-Kayan movements towards the archipelago. These beliefs in the reality of actual divine intervention in the military affairs of man would seem to have totally gripped the warriors' minds since prehistoric times to the present day.

The roots of these beliefs in the war deities are undoubtedly very ancient indeed, but it may come as a surprise to many to find that as late as the fifteenth century we can find charts outlining the risks faced by fighting men and methods of seeking divine protection from injury and defeat set out for the German noblemen in a form very similar indeed to such divination commonly employed by the Japanese heihō-sha at exactly the same time. Additionally, whilst the Japanese bushi commander called on Marishi-ten (Mārīcī), his German counterpart invoked the aid of Mary, the mother of Jesus.[2] Is this a mere coincidence? It is an intriguing point that the similarities between the late-medieval German beliefs in the metaphysical aspect of warfare and that of the Chinese and Japanese adherence to the ubiquitous female war deity are so similar as to suggest the possibility of some common and comparatively recent source; certainly one dating from sometime in the first millennium of the present Christian era.[3]

Hans Talhoffer, in his manuscript written in 1443, evidently follows the teachings handed down by the great *fechtmeister*, Johann Liechtenauer, who probably flourished in the first half of the fourteenth century. He sets out in detail a system of charts and tables whereby a nobleman could calculate the possibilities of victory or defeat that he might expect in combat. These methods almost exactly parallel similar charts and tables found engraved or painted on the *gunbai uchiwa*, war fans, carried by general commanders in the medieval armies of Japan during the Muromachi period.[4] However, there was a contrary opinion, as we have already pointed out. In a *waka* poem, a form of easily memorized instruction commonly found in Japanese heihō and the ko-ryū, Yamamoto Kansuke quotes from what might well have been the Kyō-ryū-no-heihō tradition:

> If a day's omens are inauspicious,
> They shall be poor for you both –
> Do not concern yourself with such things.
> Devote your whole mind instead
> To your efforts and training.[5]

The Manuals of the Arts of Warfare

All authorities writing about the development of the 'martial arts' in Japan are quite happy to accept the influence of the Chinese philosopher Sun Tzu and the writings of other ancient commentators, some of whose theories also date back two-and-a-half millennia, but when we take a broader view of this subject we soon realize that China was only part of the picture. It is true to say that China may have had the greatest formative influence on Japanese military thinking but the subject is far broader and more complex than that.

In my own analysis of the background development of Bujutsu[6] I suggested that the German swordmaster, Liechtenauer, may have even met and studied with one or more Japanese kenshi during one of the intermittent visits by the 'tally fleet' that sailed from the archipelago to northern China in the Muromachi period. The basis of my argument is the remarkable similarity in the techniques demonstrated in the use of the longsword and the polearms illustrated by Talhoffer in the mid-fifteenth century and those still extant from the same period in some ko-ryū bujutsu. These close similarities cannot have appeared by accident for a number of reasons and the only explanation is some sort of contact *sustained over a long period of time* which had to have been, at the very least, a year or so, but probably for far longer for the teaching and the movements to have been so thoroughly absorbed. Add to that the highly-developed metaphysical aspect illustrated by the European charts and their close counterparts in the Far East and we can argue for the clear need for much further reseach to be done on this subject.

At the time that Talhoffer produced his first two manuscripts (dated 1443 and 1459) military understanding was still firmly in the hands of the nobility and his *fechtbücher* were written exclusively for the instruction of his patron. As we have already pointed out, these were matters of great secrecy and understanding these martial principles a matter solely for the eyes of rulers or their trusted general commanders. In Japan, these conditions obtained right up to the end of the sixteenth century or even later. It was only when the iron rule of the Tokugawa *bakufu* was imposed that a profound change manifested itself in the bugei and the manuals of warfare became elevated to an almost abstract level replete with Buddhist and Confucian philosophy amongst others. Talhoffer's final manuscript, the best known of the three and written and illustrated in 1467, was almost the last 'private' treatise in medieval Europe. Practically all subsequent treatises produced in Europe in the following century reached a wider readership, albeit one probably limited to the ruling aristocracy; in Japan, with a few exceptions, such 'manuals' as the tora-no-maki now extant were not written, so far as we know, until towards the middle of the sixteenth century and even then kept under close wraps by the various ko-ryū headmasters.

The Developments in Europe

Up to the close of the fourteenth century we can only catch fleeting glimpses of regular teaching of battle technique or generalship, although we realize implicitly that this must have already been in existence for a long period of time. One of the earliest examples, now unfortunately much truncated through the loss of what may have been several sections, is the *Fechtbuch of the Secretary of the Bishop of Wurtzburg* in the Royal Armouries of the Tower of London.[7] The surviving section deals solely with the single-handed use of the broad sword and target shield but does not help other than to demonstrate the wide foot postures that may have been general in Germany at the beginning of the thirteenth century. Of interest further than the postures illustrated in this manuscript are the costumes worm by the combatants which have a distinctly archaic character, even so far as late 'Saxon' apparel shown in some earlier iconography.

Whilst most historians are only examining the methods of handling the weapons of the period, here we are seeking any indications of transmitted teaching from master to pupil that may demonstrate that these techniques and theories may have been systematically studied in medieval Europe. The first name that we have is that of Johann Liechtenauer who was active in the first half or middle of the fourteenth century and probably resided in the Nuremberg region of Swabia.[8] It is thought that this master died well before 1389 when another *fechtmeister*, a priest named Hanko Pfaffen Döbringer, wrote his parchment manuscript in 1389 now preserved in Nuremberg.[9] In this manuscript, Döbringer records every aspect of the teachings of Liechtenauer and assumes throughout the rôle of a respectful disciple, taking every opportunity to honour his master's authority on the subject. It is the influence of Liechtenauer that dominates most German, Burgundian, French and Italian writings on the subject right through the fifteenth and into the sixteenth century, but as yet we do not have a single piece of writing preserved from the master's own hand. We know little of this man and modern research in Swabia has produced only oblique evidence connecting the name to possible descendants in Augsberg or Ingolstadt.

Early in the Nuremberg manuscript is Döbringer's appraisal of the master which reads as follows:

> Hje hebt an meister lichtenawers kunst fes fechtens mit deme swerte czu fusse vnd czu rosse/blos vnd yn harnusche/Vnd vor allen dingen vnd sachen saltu merken-vnd wissen/das nur eyne kunst ist des grunt vnd kern aller kunsten des fechtens/vnd dy hat meister. Lichtenawer/ ganz. Vertik.vnd gerecht gehabt.vnd gekunst/Nicht das her sy selber habe funden vnd irdocht/als vor ist geschraben/Sonder/her hat manche laut/durchfaren vnd gesucht/durch.der selben rechtvertigne vnd worhaftigen kunst wille/das her dy is invaren vnd. wissen wolde/ind dy selbe kunst ist ernst.gancz und rechtvertik/.vnd get of das aller neheste.vnd kortzste/slecht vnd gerade czu/.[10]
>
> (Here begins Master Liechtenauer's 'Art of Fencing' with the sword, on foot and on horseback, with or without armour. And above all, one must realize that there is only one art of the sword, which may have been invented many years ago, and this is the centre of all methods, which Master Liechtenauer had mastered. As has been said before, he himself had not developed it. But he travelled far and wide and searched everywhere for information concerning the art of fighting because he wanted to know all there is to know about this serious and proper skill.)[11]

Döbringer was clearly an intelligent and well-educated man according to Wierschin, and this is shown by his skill and unusual artistry, the use of language and expression in recording the rules of fighting. He has an erudite touch and a talent for writing rhymes, yet, at all times, he speaks of Liechtenauer with deference and respect, puts his comments in the service and words of his master, and always remains as Liechtenaeur's medium. To anyone with a special familiarity with the old Japanese bugei traditions the above observations may come as no surprise; and yet we should be astonished since we are considering here a European phenomenon of the High Middle Ages and not the fundamental philosophy of master/pupil relationships that was commonplace in eastern Asia. Döbringer's evident respect has about it something that is entirely in accord with the Chinese Confucian doctrine of *li-i*, (Jap: *reigi*), respect and propriety under the rule of law. It is the simple statement of the attitude that he held towards the swordsmaster, who *may* have been removed from him by one or even two intervening teachers. Wierschin observes that Liechtenauer was a widely-travelled and 'experienced' man who dominated the teaching of the fighting arts, having a superior style, understanding in detail all its stratagems and 'secrets'. Liechtenauer always perceived mortal combat from the point of view of art. He must have been the type of master who saw in educated and able men like Hanko Döbringer pupils to be guided in such a way that their own personalities modestly withdrew behind those of the master.[12] It was this same deeply-personal moral relationship in Japan that enabled the direct inspiration of the original creator to be handed down intact for generations, even though some of these successors introduced their own interpretations to the general concepts and developed further distinctive traditions of their own as branches of the main stem. We are unfortunate that European circumstances so radically changed in the traumatic transition from the stability of the fifteenth century to the 'modernization' that thrust everything out of its path in the sixteenth; in this case where the four or five generations of successive masters who preserved and further rationalized Liechtenauer's insight finally succumbed to changes in weapon fashion that came with the late Renaissance.

While we do not possess any actual writings of Johann Liechtenauer today, we do have a number of manuscripts, several of then illustrated rather in the manner of the Japanese *tora-no-maki*, or *maki-mono*, that follow on from Hanko Döbringer's work of 1389 and particularly forming the three separate works of Hans Talhoffer, that show us quite clearly that Liechtenauer dressed his teaching in 'dark' and 'concealed' words that are only understandable to the initiated and which encapsulate the art of war. Like the later Japanese kenshi, Liechtenauer and the other *fechtmeister* saw the arts of warfare best explained from the basis of swordsmanship, developing the underlying principles from relatively simple postures and forms towards the more difficult and complex. In the later stages various other modes of warfare were introduced when the student had established a sound basis in these principles. Just as in Japan where the main weapons of the actual battlefield, before the advent of gunnery, were the bow and the spear both on foot or from horseback, it was the use of the sword that was the best tool by which men could reach understanding. We must remember an added factor that particularly applied in Europe where the patrons of these fechtmeister were generally of gentle birth and inceasingly regarded the sword and the heavy lance as knightly weapons. Only the once humble poleaxe, amongst other hand weapons, was elevated to this knightly status.

Martin Wierschin, in his study of the Liechtenauer *fechtkunst*, does not see in the many swordmasters who are recorded in the fifteenth and sixteenth centuries any sign of a direct line of descent in the first master's tradition in the Japanese sense, and this despite the use of the master's name to give some weight to claims of such descent. This use of a famous historical name was a well-tried device to gain credence that was also common in Japan where numbers of minor ryū claimed descent from famous bushi, Minamoto-no-Yoshitsune in particular. Hanko Döbringer mentions nothing of a 'circle' around Liechtenauer although on page forty-three of the Nuremberg manuscript he names other masters who supposedly, like himself, had a master/pupil relationship with the founder, without a 'circle' having existed: *'Hie habt sich an.der ander meister gefechte + Andres Juden. Josts. von der nyssen. Niclas prewssen/. . .'* However, Döbringer and the author of another very full text with glosses, Sigmund Ringeck,[13] who is described as a 'patron', but who was *Schirmmeister* of the Bavarian Count Albrecht II of Baiern-Munchen, 1438–60, clearly considered themselves as following in the line and teaching of the founder and originator since they both make use of what would appear to be original texts that they have to hand and have only added their own glosses where these texts are not clear or incomplete. The three manuscripts of Hans Talhoffer[14] are also very much derived from the same tradition but their importance lies in the fact that they are fully illustrated and span twenty-four years in this particular master's teaching career. Talhoffer was retained by the Junker Leutold von Königsegg, a feudatory of Count Eberhard the Bearded, of Württemberg, and his portrait appears in all three manuscripts with or without his lord.

On balance, while we can agree with Wierschin's opinion that there may not have been a ring of students about Liechtenauer, it is undeniable that he undoubtedly deeply influenced others with the validity of his lifetime's study, enough to inspire the priest Hanko Döbringer to commit to paper everything he could, to cause Sigmund Ringeck also to write down the tradition, adding his own observations quite distinct from those of the master, and to urge Hans Talhoffer to commit all his knowledge to three fine illustrated works. If the supposition is correct that Liechtenauer was active in the early- or mid-decades of the fourteenth century then we have a reasonable basis for saying that his original creative spirit was still extant more than one hundred years later.

Another profusely illustrated manuscript on the art of war that we should consider at this point, in examining the similarities and parallels with the slightly later Japanese tora-no-maki, is the *Flos Duellatorum*. This Italian manual of warfare was written by a Florentine master-at-arms, Fiore da Premeriacco, for his lord, Nicolo III d'Este, Marquess of Ferrara, in 1410.[15] It is illustrated throughout with ink line drawings depicting pairs of figures, variously armed and largely in combat. The principle figure in each pair, equivalent to 'tori' in Japanese terms, is distinguished either by a gold crown worn over his armour helmet or with a leg 'garter' picked out in gold just below one knee. The illustrations cover various aspects of the art of war though by no means as exhaustively as the later works by Talhoffer. Fiore da Premeriacco's understanding of these arts may have been influenced by Liechtenauer but more probably by Hanko Döbringer in his manuscript of 1396.

The *Flos Duellatorum* is important in that the line illustrations are realistically executed with care for such details as suggesting proper posture, movement, and particularly the handholds with the various weapons. There is a consistency throughout the whole work. In the main there are four pairs of drawings to each page, each pair of figures accompanied by an explanatory verse. On the surface, the primary interest lies in the seeming objective of the *fechtmeister* to depict the handling of weapons, however this seems to be gilding the lily since the whole work was intended to be a manual of instruction, or *aide-memoire*, for a nobleman already brought up to master all weapons proper to his exalted station in life, both for use on foot or on horseback. Exactly the same comment applies to all the other *fechtbuch* as well as those illustrated scrolls from Japan. While these illustrated works are of interest to modern scholars enquiring about the weaponed arts of the Middle Ages, this clearly was not their original purpose and, like their Japanese counterparts, they were intended to sharpen the recipient's understanding of the wider arts of the battlefield thereby providing their house or clan with a military advantage over their enemies.

It is to the less obvious details that we should direct our attention; for example the 'point of looking' or 'concentration', (Japanese: *metsuke*), and whether or not we can detect control exerted on the opponent through the weapon (*ken-sen-no-sen*). There are definitely some most interesting drawings in the manuscript from the point of view of our present study, ones which suggest a deeper understanding of these points, particularly in the use of the spear rather than the poleaxe or war-hammer, but whether we can deduce from this an understanding of the principles of warfare that are clear in the Chinese and Japanese forms we cannot be certain.

In the *Flos Duellatorum*, taking the use of the spear or poleaxe as our model, the artist has drawn these weapons as quite short and without the amount of detail included by Talhoffer fifty years later. In some ways, one feels that he has drawn practice weapons which, of course, might have been the case. One of the woodcuts from the hand of the German master, Hans Burgkmaier, engraved at the beginning of the sixteenth century in his series of illustrations showing the life and training of the young 'White King', Maximillian, depicts a scene within a castle practice hall which includes a number of such practice weapons.[16] Master Fiore, who drew the illustrations or supervised their preparation, aimed at simplicity but at the same time must have felt the need to include more than one drawing to express graphically what he wished to convey. This may have been the case in the *'Wurtsburg Fechtbuch'* referred to above and was certainly so in Talhoffer's work in the 'forms' pertaining to the longsword and the 'Lucern hammer'. We may be able to say, therefore, that the European viewpoint in these *fechtbuch* was to slightly amplify the master's message by incorporating the 'second phase' of a form in his *aide-memoire* drawings whereas the *bugei* counterparts were restricted to a single illustration.

The necessity for this European 'amplification' may have been due to the comparative shallowness of the contact relationship between the *fechtmeister* and his patron. In Japan, the relationship was usually far closer and understanding of the theory much deeper.

Again, it is necessary to emphasize that in both the Far East and medieval Europe such martial instruction was largely oral and between master and his pupils only. In both regions there were implicit prohibitions against further dissemination of these teachings. Some Japanese illustrations took the form of simple representations of kamae, others went further and suggested within the drawing a number of possibilities in applying the particular principle. The bugei headmaster used these graphic images to stimulate his *deshi's* imagination but not to instruct. In the European counterparts, the same may have applied but the masters found it difficult to do this in one drawing that might – and is – interpreted as a frozen moment of time. It is only in modern manuals where the objective is 'how to do' rather than 'what to do next' that we find a trend towards multiple and sometimes complex graphic presentation. In these European *fechtbuch* we see the cartoon sketch approach to the problem in its earliest stage, but we are aware that the intentions of the early *fechtmeister* may have been quite different to their later counterparts who saw these 'curious' antique forms solely in terms of combat technique.

The hand-grips on the various weapons pay close attention to the principles of what the Japanese kenshi termed *te-no-uchi* and are a significant detail as they imply that attaining mastery of the weapon's manipulation would further allow the study of its use in orthodox and unorthodox tactics. Students are encouraged by masters of the classical bujutsu to train themselves in the correct handholds as these are fundamental to progress in these arts. It is reasonable to suppose that the same considerations prevailed in late-medieval Europe to judge by these surviving *fechtbuch* and despite a lapse of nearly five-and-a-half centuries.

Master Fiore always shows the leading hand grasping the spear shaft with a relatively light underhand grip, the shaft mostly passing across the palm above the base of the thumb muscle, and gripped by the little and third fingers only, but with a slight pressure from the outside of the first knuckle of the thumb and third phalange of the forefinger. This is precisely how the same grip is made in ko-bujutsu. It is of no significance if it is the right or the left hand in the lead. The rear hand holds firmly with an overhand grip. It is considered in Japanese practice that such correct handgrips are conducive to balance and harmony; too strong a grip, especially a 'fist-like' hold, produces imbalance and therefore inefficiency.

The Italian drawings also show that Master Fiore was familiar with sliding the spear shaft through the leading hand which means that for the spears, at least, the shafts were smooth for most of their length and round or oval in section. In the Talhoffer drawings we do not see sliding but we do see a preference in the master, himself, to a double overhand grip that is also customary in classical Bojutsu and frequent changes of grip implying flexibility in use. The various kamae, postures, drawn in Master Fiore's manuscript are clearly highly developed and particularly interesting in the use of deceptive 'mist' positions where the opponent will find it difficult to 'shape' his adversary. It is a pity that the polearm techniques on foot are limited to just twenty-three drawings.

Comparisons

There are a number of intriguing questions that come from the known texts and iconography of the Liechtenauer-derived *fechtbuch*, even if we reserve judgement on the *Flos Duellatorum* because of its clear predilection towards combat of an 'Italian' nature.[17] We know that Talhoffer's three works were produced to instruct his lord in the requirements of judicial combat and the text of the first two manuscripts confirms this, but the scope of these drawings is clearly much broader. The metaphysical formulae and tables in the 1443 manuscript are intended just as much to determine the results of judicial combat and offer 'protection' as they are for the field of battle. This aspect holds just as true for Japan around this period where the kenshi commonly indulged in what amounts to judicial combat, often under special 'rules' and certainly appealed to the deities for their assistance.

Perhaps the most intriguing question concerns the many fundamental similarities in the 'forms' to be found in the Talhoffer corpus and a number of 'living' ko-ryū that have survived to the present day. From my certain knowledge and experience of part of ko-ryū bujutsu, I can assert that such forms as we are discussing simply do not occur by accident and only come from hard-won experience. To encounter them first hand in one culture like Japan is astonishing enough; to find their almost precise counterparts in a far removed and entirely different setting that dates from almost the same century span is, on the face of it, truly inexplicable!

It is difficult to escape the conclusion that there was some actual contact between Johann Liechtenauer on the one hand and a Japanese master of heihō on the other, and that this contact lasted over a sufficient length of time for the thorough understanding both of the principles and the techniques illustrating them to be assimilated in whichever direction was necessary.[18] This observation brings us back to the subject of the tengu figures found in the tora-no-maki since these figures are, in the main, always associated with the inner secrets of strategy.

Tengu Iconography

There are frequent references in the traditions to these curious 'messengers' of the war deity Marishi-ten, but most authorities in discussing the historical aspects of classical bujutsu dismiss the tengu as either an irrelevant curiosity, a superstitious nonsense, relegate them to a short footnote, or ignore them altogether. While a broad survey of the tengu tradition is to be found in *Rediscovering Budo* (Global Oriental, 2004), a paragraph or two might be of value here.

Briefly, there were two main types of these therianthropic creatures:[19] the *dai-tengu* and the *shō-tengu*. Japanese sources, when recounting the tengu legends or the many anecdotes about them that survive, assert that the dai-tengu form the senior level and the shō-tengu the lower level. However, there are several votive images of whole groups of tengu that appear on the sacred *honzon* as aides to visualizing in prayers offered to deities such as Fudō Myō-ō and, possibly Aizen and Zao-gongen, where two even more powerful or higher ranking *dai-dai-tengu* rule over the others. The affix '*dai-dai*' would accord with some upper level Shugendō ranking titles as, for example *dai-dai-sendatsu*. One of these 'senior' figures is human-faced and associated with the sun disc; the other is a beaked figure, evidently a shō-tengu and associated with the lozenge moon symbol. Both sun and moon are of importance to the Shugendō yamabushi and the followers of Marishi-ten, and originating in Inner Asia if not further west.

Dai-tengu are depicted in several *denshō* scrolls with fierce human faces. They sometimes have clawed 'bird's feet' and at others, human-toed feet. Often they have long unbound hair and hirsute faces. They may or may not be winged. If not depicted in armour, they often, but not always, wear the distinctive garb of the yamabushi. The *Shō-tengu* are similar in most respects but are nearly always winged. Shō-tengu all have beaked bird-heads, mostly feathered but sometimes more human-like. Occasionally the former are visualized as slightly larger than the latter group but there appears to be no hard and fast tradition on this point. The main characteristic is that all tengu are closely associated with the Shugendō yamabushi of at least the *sendatsu* initiated level and are commonly drawn wearing yamabushi costume, either in part or in full, and carrying objects that are also considered integral to this sect.

As one might expect, following the established custom of the martial entities, the tengu figures appearing in the denshō scrolls generally take on the rôle of *uchidachi*, the one who initiates the 'attack' in each particular form. An exception is in a tora-no-maki of the Shingan-ryu where two winged tengu, one 'dai' and the other 'shō', oppose each other. Both drawings depict quite difficult postures; one in *migi-hara-tombo*, the other in *katate-kasumi-jōdan*.[20] Their position as uchidachi accords with the fact that these creatures are the 'messengers' of the female war deity, Marishi-ten, and establishes their purpose of instructing the kenshi in the secret teachings vouchsafed by the goddess. A further exception is in the denshō of the Taisha-ryū, discussed below, where the tengu rôle is *shidachi*.

One interesting point arises from the two different levels amongst the tengu. If, as it has been suggested, their remote origins are to be found in a specialist warrior-shaman tribe who accompanied one or other of the Puyŏ-Kayan usurper groups and subsequently retired into certain mountain districts to later re-emerge as the proto-yamabushi, then it might be arguable that the two divisions represent, originally, not one but two warrior-shaman tribal groups. Be that as it may, one characteristic of the dai-tengu that is most noticeable is how remarkably 'European' or 'Western' are their facial features. Japanese folk images of the tengu, commonly to be found all over the country in many large or small Shintō shrines and Buddhist temples always give these creatures comically long noses,[21] but the human-faced warrior tengu of the denshō, whilst displaying long noses are much closer to the native descriptions of Europeans and their depiction almost from their first setting foot on Japanese soil in 1543.

If, as has been mooted, one of the first warrior contacts between the bugei tradition and a European master swordsman, came with the arrival of Johann Liechtenauer at one of the ports used by the Japanese 'tally fleet' in the early fourteenth century, who better than this German *fechtmeister* to fill the rôle of a tengu emissary of Marishi-ten? Perhaps Liechtenauer, evidently a man of fairly mature years though still filled with the necessary energy to complete his quest to discover all that was to be found about swordsmanship, had finally reached the northeastern Chinese coast gaunt and drawn in his face; a very different sort of figure from the smaller round-faced Chinese and Japanese? Undoubtedly, such a man, skilled in the arts of war, would have impressed most warriors that he encountered where an exchange of ideas was possible rather than mere brutal combat. Extending this speculation further, to the possibility of such a meeting having taken place, the first important tradition known to have had *tengu-shō* 'secret' levels in its *okuden* was the Kage-no-ryū founded by Aisu Ikkyō around the beginning of the Sengoku period, say 1475–80. We do not know who this swordsman's antecedents were but we do have a strange account of him meeting with an 'avatar' who had connections with China.[22] Such anecdotal glimpses of old events abound in the bujutsu traditions and very possibly this is an echo of the meeting of a kenshi with Liechtenauer over a hundred years earlier. But one thing it is true to say is that in the German corpus of sword and spear use drawn by Hans Talhoffer there is no discernable sign of 'Chinese' influences, only the already-noted uncanny closeness of some of his forms with many from the oldest and still uncorrupted kata traditions in Japan.

Unfortunately, the Kage-no-ryū *tora-no-maki* have not survived. Still extant, however, are illustrated scrolls from two related traditions that descended directly from the time of the third successor of the Kage-no-ryū, Kamiidzumi Ise-no-kami Nobutsuna, a man who probably actually trained under the founder, Aisu Ikkyō. Later changing his name to Hidetsuna, Nobutsuna had devised the Shinkage-ryū which flourished and became extremely influential under the aegis of the Tokugawa-ke after Sekigahara. A second student, Marume Kurandō-dayū (b. 1533), founded the Taisha-ryū-no-heihō, a battlefield system that unfortunately did not fully survive Bakumatsu in its complete form but does remain in a fragmentary condition to intrigue Kendō historians with some highly original forms.

The Yagyū Shinkage-ryū tengu drawings are contained in a scroll that was presented to a Nō actor by Matsudaira Nobusada in 1707[23] but are believed to be based on illustrations dating from the middle of the sixteenth century. These illustrations of the *Tengu-shō kuden* are thought by some authorities to have come directly from the Kage-no-ryū but it is always possible that Aisu Ikkyō only transmitted his tengu levels by word of mouth and committed nothing to paper. The second group of tengu drawings are contained in a series of denshō given to a Nabeshima-han samurai, one Ōbushi Kansuke, a deshi of the Taisha-ryū around 1675. Both sets of illustrations are exceptionally well-realized and are probably by the hand of a professional artist working under the direction of the ryū masters but from the importance of these secret teaching levels it is more than likely that the drawings were directed by the incumbent headmaster, personally. Despite their comparative late date, these tengu are thoroughly 'traditional' and show nothing of the decline towards folk art that we often find in later-Edo-period budō illustration work.

Yagyū Shinkage-ryū Tengu

These are eight in number and all depict individually named dai-tengu with bird-clawed feet. Each of the tengu wears the ubiquitous *tōkin*, the small eight-sided pillbox hat of the yamabushi; *sashinuki-bakama*, a kind of 'plus-fours' trouser; *suneate* shin-armour; and one carries a feathered *tessen* fan on his back. Significantly, this type of feathered *tessen* is always a symbol of invisibility. All eight tengu have *tachi*, the slung swords worn by bushi from the Heian to the end of the Muromachi period but considered extremely unusual in Edo times.

We can learn little from these tengu other than in the main they are shown with deep postures and all have drawn swords, four in variations of *seigan-no-kamae*, left or right foot leading, and four in *gedan-no-kamae*, three right and one left foot leading. Matsudaira Nobusada brushed in a brief description of each form but, as with many such glosses in these traditions, these comments would only be understood by a deshi who had reached this level and are meaningless to all others. The comment brushed at the end of each gloss tersely states: 'an oral transmission'; it is one found in a variety of forms throughout the bugei denshō. One point of interest, despite this, is the characteristic very wide *hanmi* postures shown in many of the kamae.

A last curious point is that all the tengu are given proper names, perhaps suggesting that much earlier in the Shinkage-ryū development, possibly as far back as the time of Aisu Ikkyō, certain principles were thought to be transmitted by Marishi-ten in the charge of different tengu messengers. In no particular order, these names and their associated forms are:

Kōrin-bō – Rankō;   **Chiraten-(bō)** – Koradome;   **Fūgen-bō** – Noritachi; **Konpira-bō** – In-no-kasumi or Hashigaeshi or Tōtōkiri; **Eii-bō** – Kiritsume; **Tarō-bō** – Komurakumo;   **Karan-bō** – Subete-koran-uchimonodome; **Shutoku-bō** – Kissakizume.

Taisha-ryū Tengu

Of much greater value from the point of view of stimulating the intuitive mind of a swordsman whose understanding has already been broadened by experience are the beautifully realized illustrations in the Taisha-ryū denshō in the possession of the Saga Daigakuen.[24] Once again, nothing is described in detail in any of the brief texts accompanying these tengu and human figures, but the ryū had commissioned the illustrators to suggest various possibilities that will follow from several postures in the course of subsequent action. Whilst one can never be sure of the precise truth without study within the tradition, nonetheless the 'frozen' action drawn in these spirited pictures strikes a creative chord and might even be developed into more tangible form. Naturally, during the late-medieval period such 'internal' documents would have remained strictly within the swordsman's possession or be passed on within his family archives, but the fact that a master swordsman from another tradition could make deductions from this visual material underscores the extremely sensitive nature of all these advanced forms within heihō. It is only now that we are able to study these entities but the tragedy is that so much has been lost during the 'modernization' of Japan after Bakumatsu.

While it is possible that all the denshō given to the Nabeshima-han samurai, Ōbushi Kansuke, in the seventeenth century, may not have been seen at Saga University, those that were examined depict no less that ten tengu. Additionally, there are three kenshi figures which, unlike the Yagyū Shinkage-ryū human swordsmen, are wearing quite normal apparel. The first point of interest is that all these figures, tengu or human, are in postures suggesting that they are about to draw their long swords in the manner of Battō-jutsu. Not only that, each figure is shown with the sheathed blade edge downwards so that whatever technique is contemplated it will be 'earth to sky'.

Whilst the Taisha-ryū may have developed out of the Shinkage-ryū of Kamiidzumi Ise-no-kami Hidetsuna, the former in its turn came from the Kage-no-ryū. This allows us to examine all these tengu forms from both descending transmissions. It is also fortunate that the Taisha-ryū is thought to have exerted a considerable influence on those systems of Iai-jutsu (Battō-jutsu) inspired by Hayashizaki Jinsuke, a contemporary of Marume Kurandō-dayū.

There seems little doubt that these tengu-shō levels within the Taisha-ryū are based on the principles of heihō that were contained in Aisu Ikkyō's Kage-no-ryū. As we have already seen, most of the Drawing-sword techniques discussed fall into the category of 'unorthodox' or 'ch'i' tactics. The illustrations all depict tachi-mounted longswords that are worn thrust through the figure's *obi*, or waistband. All the forms vary from partly flexed knee postures to quite deep ones and like those drawn in the Yagyū Shinkage-ryū denshō, the rear foot is in each case well-turned back. This suggests to bugeisha that the underlying concepts are 'old-fashioned' and derive from the first half of the Muromachi period despite appearing in late-Muromachi ryū. Nine of the tengu are dressed fully as yam-abushi and all are winged. Six are shō-tengu while the remaining four are dai-tengu. In conformity to the later Hayashizaki Jinsuke ryū-ha, these Taisha-ryū tengu are all wearing tiger skin *suneate* or *kyahan* on their shins, these Iai-jutsu systems having a distinct predilection for tiger imagery or metaphor in their teaching levels, but only one of the figures is given a tiger skin sheath to his sword. (The Shinkage-ryū tengu wear old-fashioned plate *suneate*.)[25]

In front or superimposed on each tengu and two of the human kenshi, though not grasped by any figure's hands, are drawn swords in various positions. These swords are placed in a manner difficult to evaluate without direct experience of the inner teachings but, as commented above, in several cases they do stimulate a creative response. It is possible that the ryū *sōke*, or headmaster, intended to remind Ōbushi Kansuke of sequences of movements within certain forms or even within a group of related forms illustrating a developing principle. One of the tengu forms suggests that the sword is drawn to counter restraining attacks attempted with *mojiri* or direct spear thrusts aimed at the torso – with a strong upwards draw followed by a rapid downwards cut. Two more suggest that the draw threatens the attacker and allows the kenshi to circle back past his left shoulder and cut down stepping in. This type of move we have already encountered in the discussion centred on spiralling techniques.

Other Tengu Illustrations

Quite a number of ryū denshō contain tengu figures, either associated with the secret *kuden* levels, like the beautifully-illustrated four panels in the *Tengu Hidenshō* drawn by or at the behest of Shimada Hideyoshi-zō, or singly as part of the preamble to a tora-no-maki that links the transmission with Marishi-ten as in the Musō Kenshin-ryū denshō, in the Eihiko Reisenji collection. In this latter case, the figure of Marishi-ten rides on the back of a charging *inoshishi*, or wild boar, armed with her usual weapons of drawn bow, jumonji-yari, sword, *tessen* fan, and what may be a vajra-hilted blade or baton. Ahead and facing her is a dai-tengu dressed in the yamabushi manner, right hand grasping the hilt of his tachi sword as though about to draw it upwards like the tengu in the Taisha-ryū drawings. The sword has a tiger's tail scabbard and this tengu's suneate or kyahan are also tiger striped. The winged tengu's feet are human but clawed. Both figures follow Buddhist convention in depicting certain fierce deities and have upstanding hair of the type often seen in the iconography of Aizen and Zao-gongen.

Finally, the still flourishing Tenshin Shoden Katori Shintō-ryū, whose origins can be traced back to the close of the fifteenth century, has many forms of heihō directly adhering to transmissions believed to derive directly from Marishi-ten. A number of techniques utilizing the *bō, naginata, tachi* and *ōdachi* display either *karasu-tobi* (crow-hopping) or *tengu-tobi*, sudden unorthodox moves just prior to delivering a devastating attack. The tora-no-maki of this ryū includes an illustration of a three-headed Marishi-ten balancing on one leg on a charging inoshishi. She holds all the usual symbols of the war deity in her several arms. To reinforce the link with the Buddhist mikkyō, the upper part of her body is framed in the cosmic flames usually associated with Fudō Myō-ō. There can be little doubt about where this ryū believes its origins lie.

■ CHAPTER 14

Conclusions

With so much of the warrior tradition lost in the course of the last century-and-a-quarter we are severely restricted in making enquiries directly from the living traditions of the bugei as it still existed at the end of Bakumatsu. In those old bujutsu entities that survive from origins as far back as the Sengoku period in the late-Muromachi, to be able to really understand the significance of the teaching a student must devote more than half a lifetime and totally immerse himself within and under the restrictions imposed by the ryū. For a non-Japanese to do this is difficult in the extreme, and that is not to say that it is any easier for one of Japanese birth and ancestry. The integrity of these warrior systems depends on secrecy in the sense that was envisaged by Sun Tzu so long ago.

At the beginning of the last century one European, Willem de Visser, a man steeped in the folklore of both China and Japan, made a deep study of the tengu which must have occupied him for a number of years. At that time many sources still existed and yet he almost totally missed the tengu tradition within the bugei and the two main accounts that he cites, those of the associations with military training and Minamoto-no-Yoshitsune and a deeply interesting incident at the beginning of the Nanbokuchō struggles, he failed to follow up, thus depriving scholars today of what must have been many, many 'living' traditions amongst swordsmen and former samurai. De Visser's accounts, whilst of great interest, simply bring together the folklore and the rather negative but successful Buddhist 'demonization' of the tengu that attempted to relegate them to the rôle of mischievous forest 'goblins'. A number of modern studies published over the last decade or so have tended to perpetuate this restricted viewpoint and this is understandable when Western commentators in the field of the so-called 'martial arts' have ventured to air their varying interpretations based on minimal real experience of the nature of ko-bujutsu and allowed themselves to be seduced by the watered-down Edō-period viewpoint at best or the 'popular' superficial definitions of the media or the film entertainment industry at worst. 'A little knowledge is a dangerous thing' was a well-known saying in my youth and how true it is in matters so important as warfare.

The Japanese masters of the bugei and certainly those who were the creators of the first systematic rationalization following the clear guidelines set out nearly two thousand years before them by such inspired Chinese philosophers as Sun Tzu and his descendant, Sun Pin, were rational pragmatists. They were men whose very lives depended on their abilities. These were not warriors who sat *zazen*, mumbling interminable prayers and were struck smartly on their shoulders when they fell asleep, in order to reach some sort of understanding. Their descendants, enjoying the peace of the Tokugawa Bakufu were able to do this and dress up their theories with liberal dollops of Zen and Neo-Confucian semantic prosy talk, but it was those men who fought and survived the bloodbaths of the Sengoku period who clearly subscribed to the notion that in matters of intuitive action that they were unable to express in so many words, their brilliant insight into the principles of *successful* warfare was derived from Marishi-ten and came to them by means of these same tengu. Not the tengu of a jealous and hag-ridden Buddhist prelate, fearful of any heresy, but the tengu that had, since the dim beginnings of the warrior past been present in the secret instruction held in the custodial hands of certain yamabushi. These were men who like the famous general, Toda Seigen, when asked about his philosophy, thought long and hard before replying:

'Don't die!' They were men who, quite simply, had followed the yamabushi paths.

If we totally disregard the ideological writings of the Edō-period swordsmen and concentrate on the denshō that survive from the late-Muromachi, we find nothing that is really clear; practically every single reference even to simple matters as sword postures and where the feet should be placed, is couched in language that precludes any real understanding except by an initiate. Even these initiates will change their interpretation periodically and not a single one of these aspiring masters will clarify anything to their deshi. They must find the truth out for themselves and this can only be done by wholly immersing oneself in the *oral* tradition.

The bugei are full of this obscure terminology, difficult to comprehend then but impossible now once in the hands of the Western 'adepts' who claim rank and understanding that only a handful of men in *Japanese* history have ever been considered to possess. These are 'mouth to ear' matters; absolutely nothing is what it seems and that brings us back to Sun Tzu and 'all warfare is based on deception'. Perhaps, if we are compelled by necessity towards Zen, we should try to understand the real meaning of 'self-delusion'?

Examples of the Terminology Used in Heihō

The hurdles set in the path of understanding are immense and, for most, almost impenetrable. Not only are many of the terms drawn from the mikkyō and then applied within the bugei but they are often variously interpreted in the Japanese language as well as into English; others are names purposely derived from the Chinese, some obliquely alluding to one or other classical themes. They may, in some cases, appear to be transparent when in fact they are truly opaque.

For this reason the expert of heihō, the master tacticians of the Muromachi studied Sun Tzu and took his words to heart. Probing the potential enemy through the use of agents or spies is essential but the wise commander also knows that the opposite is true. The enemy may certainly be employing his own agents and some of these may be long term 'sleepers', well-hidden even for many years, some gaining positions of trust

Sun Tzu wrote: 'Determine his [the enemy's] dispositions and so ascertain the field of battle.' 'Probe him and learn where his strength is abundant and where deficient.' The underlying reason why progress in ko-bujutsu was so slow and thorough was to ascertain the probity of the deshi who sought knowledge from a master. It is why some masters never took students. In an era when the use of espionage, even simple neighbourhood surveillance, both internal and external, was all-pervading, few men could really say that they trusted even their own family members. It was as the Portuguese Jesuit priest, John Rodriguez (1559–1633) once wrote, here paraphrased, that: 'The Japanese have three hearts; one that they show to their family and friends, another that they keep secret to themselves, and a third, deep inside, that they show to no one.' The arts of the bugei were, to many senior samurai, truly matters of life and death. To draw sword on someone else, also armed, often meant that victory or defeat could be measured by millimetres or a split second of time. The techniques and underlying principles studied in the privacy of a fortress yard or the secrecy of a closed dōjō area within a shrine became in the course of time the strength of a group or clan, spelling the difference on many occasions between victory or wholesale extinction for many bushi and their entire families. This was information that one could not allow to spread out beyond the control of the master or his senior initiates. But these same secret principles must be tested, hence the use of *ta-ryū* forms, often excellent in their own right, but which would give little away in the event of defeat. The importance attached to these 'rubbish' forms is fully evident in that the custom was so ingrained that they continued in use certainly to the close of the Meiji period, and echoes are to be found even in the present day.

> It is according to the shapes that I lay plans for victory but the multitude does not comprehend this. Although everyone can see the outward aspects, none understands the way in which I created victory.[1]

CONCLUSIONS

Some terms deliberately defy normal understanding as they suggest that the secrets derive from the deities, themselves. Examples relating to the names given to some traditions are:

Musō Shinden:	Transmission of a deity's dream vision.
Tenshin Shōden:	True and correct transmission from a deity.
Kashima Shinden:	True transmission from (the deity) of the Kashima Shrine.
Shinkage:	Transmission received through the influence of a deity, or Divine Shade (transmission).
Shindō Musō-ryū:	Shintō Dream-vision tradition.

Then we find terms that are sometimes descriptive, symbolic or metaphoric.

Shōtengu:	A correct transmission by means of a tengu.
Yubi-no-makura:	Pillow-dream, an inspirational vision.
Yumemakura:	Dream revelation, or Appearance in a dream.
Gomusō-no-jō:	Five Dream-vision Staff Strategies.
Okugi:	Secret Principles, or Inner Mysteries.
Gokui:	Secret Principle, or Secret Techniques.
Muniken:	Sword Strategy of Non-Duality, or Matchless Sword.
Shinmyō-ken:	Marvellous Sword Strategy.
Rei-ken-den:	Magical or Miraculous Sword Strategy.

Other terms are even more abstract:

Mu-tō:	No-sword.
Mu-nen:	No-thought.

Isshin-furan:	One Mind without Confusion.
Mu-shin:	Freedom from Subjective Thinking.
Muga:	Selflessness.
Myōyū:	Mystic, or Wondrous Activity.
Fudōshin:	Imperturbability, Selflessness.
Kuden:	Oral transmission.
Marishi-sonten-no-kuden:	Oral transmission from the Worthy Marishi-ten.
Kasumi:	Invisibility (lit. Mist).

As we have already noted, the names given to individual techniques and kata forms, whilst often colourful, are no real help at all even to the novice within the tradition. They do become meaningful, however, as the deshi reaches the upper levels, probably because of their nature as *aides-memoire* coupled with an understanding of the secret nature of the forms. The following are a tiny selection from the bugei traditions:

Shishi-funjin:	Lion-spring.
Koran-no-iri:	To gather oneself like a tiger, then attack.
Yaigasumi:	Under the cover of darkness.
Tsukimi:	Threaten the eyes.
Sarutobi:	Monkey Flying.
Tōbō:	Sword and Stick.
Tsubame-mawashi:	Swallow-dipping.
Tsubame-gaeshi:	Swallow-somersault.
Yamakage:	Mountain-shade.
Tsukikage:	Moon and Darkness.
Uranami:	Waves in a Bay.
Ukifune:	Floating Boat.
Sekkō:	Helmet Breaking.
Embiken:	Flying-swallow Sword.
Hiryuken:	Flying-dragon Sword.
Shishi-giri:	Lion Cut.
Ranken:	Wild Sword.

Eiibō:	Laying a trap.
Tensetsu-Ransetsu:	Close Circling Impetuous Cutting.
Katsunin-ken:	Life-giving Sword.
Tomoe-ken:	Commas-in-a-perfect-circle Sword.

The listing of more of these kata names would be both pointless and practically endless; they are interesting because of their inventiveness and 'colour' but they give away very little. One that is slightly less impenetrable than the rest, perhaps, is the last on the above list and illustrates rather cleverly the principle contained in the observation of a familiar symbol, that of the *mon*, or emblem, often to be seen in Shintō shrines. This is the *ni-no-tomoe* or the *mitsu-no-tomoe*, double or treble 'commas'. This suggested to the kenshi devising his form to illustrate the strategy that the sword might be drawn high to the left-front, swept round to the right and cut back again to the left. Such large movements would mislead the opponent and have a devastating effect. If this form is as old as is claimed, and there is little reason to doubt the assertion, it can certainly be dated to the second half of the fifteenth century.

This survey, short as it is, also reinforces the fact that despite Sun Tzu's apparent distaste for placing any reliance on the deities or omens, most, if not all, the medieval Japanese masters sought their enligtenment, or to put it another way, honed their understanding, by means of practising the austerities of the Buddhist or Shugendō mikkyō. Heavy emphasis was placed on the customary final period of seclusion within the grounds of a shrine or temple, sometimes lasting for more than a thousand days, praying to the deity – usually Marishi-ten – and honing the 'forms' into a satisfactory finished structure. It is important to remember that whilst many of these warriors formulated brilliantly original kata they were not, in the main, educated, but hardy and toughened fighting men, all of whom had been through the baptism of fire on the battlefield. It would be interesting to focus study on why it was that certain shrines were favoured in this final process towards perfection. There are the obvious ones like the great warrior shrines of Kashima and Katori, or the Suwa-taisha, but the selection of others, like the strange Udō Myō-jin cavern-shrine on the wild southeastern coast of Kyushu, need explanation.

However, despite Sun Tzu's strictures, there can be no doubt that the mikkyō exerted an enormous influence on the theories underlying the heihō and, therefore, on the conduct of generalship in the field. To what extent religious theology actually influenced the kenshi's minds is open to debate. There is no doubt that the bugeisha were sincere in performing the rituals demanded by Shintō at the commencement and end of practice and these rites may first have been those of Marishi-ten, but it is possible that these hardened practical warriors merely hedged their bets just in case the deities appealed to could actually offer protection in a life and death situation. What prolonged exposure to the esoteric practices really produced, and especially to those masters who followed the extreme privations inherent in the 'yamabushi ways', was a high degree of self-reliance. They were undoubtedly well-aware that such practices, helped by the conventions of focus in such deities as Fudō myō-ō or Marishi-ten, would give them full confidence in their own abilities coupled with steadfast sharpened minds used to coping, largely without support, with difficult or downright dangerous situations. This ability to stand by their own decisions was backed by their certain knowledge that they had the 'protection' of Marishi-ten.

This reliance is reflected in a *waka* dating from the early-sixteenth century and brushed by Yamamoto Kansuke for his projected *Heihō Okigishō* that clearly decries a reliance on the improbables of religion and reflects Sun Tzu:

> If a day's omens are to be bad
> They will be bad for both sides.
> Pay no concern to such things.
> Concentrate your mind instead
> On your effort and training.

Reliance on one's own ability and an implicit belief in the strength of one's master's teaching is a constantly recurring admonition in the classical bujutsu dōjō. There is no room for self-doubt or questioning why or how the forms are taught or whether they would actually work if put to the test. Such a student shows the wrong spirit and, following Confucius, should soon be shown the door. There is the famous advice given to the young Watanabe Kazuma, a retainer of the Ikeda-han by his supporter, a swordsman named Araki Matayemon, before their *katahi-uchi* fight at Sagara in Iga province. The young samurai was told that 'If his enemy cut into his skin, he must cut into his flesh; if his enemy cut his flesh, he must cut to his bones; if he was cut to his bones, he must take his enemy's life.'[2] There was no compromise but underlying this was the need to conserve energy in combat. However, Araki Matayemon's advice was already part of the heihō ethic and current at least a hundred years earlier. Yamamoto Kansuke Haruyuki, again, in discussing 'Spirit':

> Let your opponent cut your hand – you cut his neck,
> Let your opponent cut you lightly – you cut him deeply,
> Let your opponent cut your left hand – you cut his left leg.[3]

■ CHAPTER 15

'Subtle and Insubstantial . . .'

The great Takeda-han strategist, Yamamoto Kansuke, hints in a biographical note that he began to understand the real meaning of strategy when he was about thirty years of age, which gives us a date around 1525, the year that another of the famous bugeisha, Matsumoto Bizen-no-kami, was killed in a minor fight in eastern Japan. This is interesting if we accept some persistent traditions amongst swordsmen of the Kashima area of a friendly connection between the mighty Takeda and the much smaller Kashima-han. Matsumoto Bizen was the founder of the Kashima Shin'ryu-no-heihō, alternatively known as the Matsumoto Shintō-ryū, and a fellow senior councillor of the Kashima-ke along with Tsukahara Bokuden, a renowned skilful strategist. Some of the comments that we read in the *Heihō Okigishō* closely resemble teachings in Bokuden's transmission that has survived as the Kashima Shintō-ryū-no-heihō to the present day. As an example, some twenty years ago, at the end of a third short return visit to Kashima-jingu, I was advised to view all ko-bujutsu 'using my imagination'. Intelligent comparisons and evaluation were fundamental to understanding. Yamamoto Kansuke concluded his work on heihō with the words:

> Strategy is something to be done with imagination.

'SUBTLE AND INSUBSTANTIAL . . .'

Eight years before, in April 1976, after I had spent several weeks at the home of Professor Kuroki Toshihiro, he also stressed the importance of imagination in evaluating everything experienced by the kenshi. Unless a swordsman, meaning a man who was familiar with the theory of several types of bujutsu, the 'Four Pillars' of the bugei, was able to study for long periods of time under a number of masters, something that has always been nearly impossible, he would be unable to assimilate the detail of those different systems to any great extent. However, the intelligent and observant kenshi, once witnessing a demonstration of unfamiliar technique would immediately recognize something of the basic principle behind it. He would, in the words of Confucius, 'be shown one corner of the problem and be able to come back with the other three'. Kuroki sensei was pointing to Sun Tzu's injunction to 'shape' one's opponent.

Throughout this study, focusing as it does chiefly on the *ch'i* aspect of warfare, the 'unexpected' or 'unorthodox' strategies as opposed to the broader field *cheng* of orthodox strategies required in more set-piece warfare, the underlying principle understood from ancient times is 'to know and to act'. This was stated in a variety of ways by the medieval Japanese masters of the Muromachi; they well realized the importance of gaining and evaluating information of every kind, however small, and, equally, of denying such knowledge to a potential enemy. Expressed in another way, this principle was handed down to their posterity in the saying: 'First the eyes, then the feet, then the sword, and lastly the spirit.' It was a timeless truth of heihō gained on the bloody ground of the battlefield and transposed seamlessly to the softer forms of budō we enjoy to-day. It is a principle precisely contained in the maxim: 'To know and to act are one and the same', to state it in full, and reflects Sun Tzu from two-thousand-five-hundred years ago to be echoed time and again down the intervening centuries.

Iai-jutsu has been cited as the basic model here since it generally reduces military complexity in the field to its simplest form and contains techniques that are practically all 'unorthodox' in character. Many of the forms of the Muromachi battō were devised to illustrate recommended tactics to deal with relatively common everyday encounters. Some of these were discussed briefly by heihō-sha like Yamamoto Kansuke as well as being encapsulated within other ko-ryū. Of these, many were easily created forms, variations of which can probably be found in any number of traditions; for example, a practical method of dealing with a number of opponents met in a street, or how best to defend oneself when threatened from behind. Others address simple tactical situations such as combat conducted across a mountain slope or on a boat. In general, these matters were dealt with not only by battō but by kenjutsu, the use of the halberd and spear, and by archery; many being within the field of 'orthodox' warfare of any size from an individual combatant to large units of men. It was the question of sound strategies within the totally different fluid area of 'ch'i' that these expert strategists were most concerned and it was here that the utmost secrecy was mandatory.

When the political and military unrest of the Sengoku period finally ended, the *raison d'être* that produced the fighting arts of the warrior partially ceased to exist, the latter-day samurai were loth to give up their patrimony and flocked to the bourgeoning budō teachers who now became rather fashionable as they no longer always had to give instruction in techniques that actually worked. Accounts abound of established 'masters' becoming jealous of a number of very able kenshi who continued to study in the 'old' traditions and several ugly and discreditable incidents are recorded. The major shock came with the clever tactics and planning of the Akō retainers in the famous 'Forty-seven Rōnin Incident' of the Genroku period, at the close of the seventeenth century (1688–1703), and the traumatic upheavals of Bakumatsu when two bloody civil wars were fought pitting the last of the classical samurai against the technology, organization and increasing collective discipline of the government conscript armies. As pointed out earlier, we are extremely fortunate that some, at least, of the traditions of ko-bujutsu managed to survive all this to reach us to-day.

During the two-hundred-and-fifty years of Tokugawa rule many of the ko-ryū transmissions either petered out and were lost or metamorphosed into budō under the increasing influence of more

'SUBTLE AND INSUBSTANTIAL . . .'

fashionable philosophies and changing social attitudes, particularly the rigid dictates of the Bakufu and powerful rulers of the provincial domains. There was always a fear amongst the samurai bureaucrats that some bujutsu dōjō might become secret hotbeds of warrior intrigue, something that did in fact happen in the mid-nineteenth century. The hiatus that came after the last civil war in 1877 and the resurgence of interest in their warrior past that began to take shape at the beginning of the twentieth century meant that numbers of able former warriors, those who had survived the terrible events of the final upheaval, had been lost due to the pressing need simply to survive loss of employment, and budō was altered to conform to the pressures of education and, gradually, of sport. Some of the ko-ryū transmissions continued, though, and soon encouragement in the form of shin-bujutsu and a renewed interest in orthodox Kendō practice in the Japanese Imperial Army and especially the Japanese Imperial Navy ensured that standards were set high. With this interest came a renewal of studies of Sun Tzu and the Chinese martial classics with several new translations and commentaries on the *Ping Fa* being published between the two World Wars. Despite all this, it is in the ko-ryū that the richest veins of information are to be found and we are still able to examine elements of the bugei untainted by developments that entered immediately after the sengoku period came to an end.

Historically, as we have seen, a later headmaster of the Hasegawa Eishin-ryū-no-heihō, as early as the late-seventeenth century, was moved to devise and insert a whole set of forms that would help his young deshi to grasp the basic principles of Iai-jutsu. This was a bare two or three generations after the time of their forefathers who had fought through the last years of the Sengoku and knew very well what they had been about. It was also a time when possibly more than a thousand ko-ryū existed and iron must still have been evident in the souls of many warriors, at least those of rank. Maybe we can follow his reasoning but the fact remains in this example that the Ōmori-ryū Iai, as it is now known, only contains one or two forms that provide a true transition from the earlier Iai-jutsu formulated and transmitted by Hayashizaki Jinsuke. To put this example of the gradual loss of understanding into a more general perspective, we must become aware that the understanding of heihō had already declined to a significant extent when the so-called Shimabara Rebellion broke out in 1637 and that it took an army of 100,000 men two full months between April and June of the following year to take Hara Castle held by around one third of their number. The Tokugawa Bakufu experienced serious problems in finding a commander with sufficient understanding of strategy to bring this sorry affair to its sanguinary conclusion.[1]

The question whether or not the Japanese understanding of heihō declined after their great wars of the Sengoku period died away to distant memory and their costly mistakes in the Far Eastern theatre between 1937 and 1945 does not concern us here. There can be little doubt that the full flowering of their interpretation of Sun Tzu and parts, at least, of the teachings of other classical Chinese military philosophers had a great impact on the bugei and must have made a significant contribution in the field of medieval conflict. Our interest, in the present day, may rest on the fact that these teachings and their applications enriched immeasurably the ko-bujutsu and later formed the foundation of the subsequent metamorphosis into classical budō, some of the teachings entering the modern shin-budō systems that are so popular to-day. For us, though, the cut-off point comes with the great battle fought at Sekigahara on 21 October 1600. The total number of combatants have been estimated at 210,000 men with 30,000 left dead after the fighting finished.

Our focus has been on the concepts of warfare that were developed during the hundred-and-fifty years preceding Sekigahara; the application of the 'unexpected' in particular. The final word should, therefore, rest with that master strategist of the Takeda-han, Yamamoto Kansuke:

> Defeating your opponent successfully means attacking before he realizes he is being attacked. Even if he understands the situation, attack him before he is able to prepare.

Notes

Chapter 1: The Muromachi Warriors and their Approach to Sun Tzu: Some Historical Benchmarks

1 There are a number of excellent surveys available, particularly Donn Draeger's trilogy, *Classical Bujutsu*, etc. For an overview, my study *Rediscovering Budo* may also be helpful.
2 It is notorious that numbers of so-called 'masters' will claim such knowledge but, in the final analysis, most will do so for commercial rewards and self-advancement. Matters have hardly changed since Miyamoto Musashi made such comment at the beginning of the seventeenth century.
3 It is partly in obedience to the spirit of this blood oath that writers who have been inducted to one or another of these ko-ryu are often reluctant to give precise descriptions to specific transmissions in their texts. Generalizations are permissable; specific references are not.
4 See Carmen Blacker, *The Catalpa Bow*, and Knutsen, *Rediscovering Budo*.
5 Yamamoto Kansuke, *Heiho Okigisho*, translated by Obata Toshihiro.
6 Naginata: a halberd or glaive – as defined by H. Russell Robinson late HM Royal Armouries.
7 The Shugendo connections continue to the present day though in a much reduced form.
8 This legend would put the original foundation at around 350 CE but the first actual building was thought to have been constructed early in the eighth century. This probably replaced a small miya, or 'palace', dating from the sixth or seventh century. The kamisama at the present shrine, moved to the centre of Kobe in late-Meiji, is female but not actually named but the shrine authorities believe she is Jingō-kōgō.
9 O-kuni-nushi ('lord of wide lands') seems to have been an hereditary title used by the Izumo rulers over several generations.
10 Take-mika-dzouchi-no-kami is also enshrined at the Kasuga-taisha in Nara, revered as the tutelary deity of the Nakatomi clan who shortly thereafter became the Fujiwara and entered the Imperial line.

NOTES

11 The Zao-dō houses a famous wooden statue of the fierce deity Zao-gongen, revered in the sangaku-shinken and Shugendo. Although Zao is found in no known sutra he is clearly based on the Five Mighty Bodhisattvas. It is thought that he originated in Tibet.
12 The Taizoku mandala is usually termed in English the 'Womb-World' mandala. The Kongo-kai mandala is that of the 'Diamond-World'. The former embodies compassion for all things as they are; the latter is the 'wisdom-body' enfolding all beings and phenomena in the Single being of Dainichi Nyōrai. (See: Yamasaki, *Shingon*, etc.)
13 The late Thor Heyerdahl thought that there was some merit in Snorri Sturlason's account in the Prose Edda that the Æsir people led by Odin had originated in Sythia before their migration west and north to Scandinavia. Part of his argument pointed to the wide prevalence of the raven imagery that occurs right across the European-Asian landmass.
14 There seem to be connections in these symbolic 'messengers' with the southeast Asian garuda and the Chinese dragon. In the latter case this may be a conflation with some indigenous myths. One error that has almost certainly crept into the academic evaluation of these bird symbols is the identification of the raven/griffin modelling found is some very early Japanese and Korean sword koshira (pommels) as being 'dragons' and therefore from Chinese influences when they clearly originate from Central Asia.
15 Shin-budo: those modern systems that were either modified from the more classical entities post-1900 (Kendo is a good example) or formulated in the second half of the twentieth century, ostensibly based on earlier principles. Whilst Judo is classified as a 'shin-budo' at least it can claim to have always been a martial sport.
16 D. T. Suzuki, *Zen and Japanese Culture*, (1997) In my own experience I have yet to come across a single aspect of classical Kendo that contains, even remotely, Zen-do, that cannot be seen in the older beliefs outlined above.
17 John Gray reviewing 'Descartes: the life of René Descartes and its place in his times', by A.C. Grayling; Free Press, 2005, in *New Statesman*, 17 October 2005.
18 This view was advanced by the Kendo historian, Yamada Jirokichi, in his *Nihon Kendo-shi*, (1919). Yamada was sometime headmaster of the Jiki-Shinkage-ryu and bequeathed to posterity a number of highly important studies connected to the ko-ryu traditions.
19 In postulating a date for the *Heihō Hidenshō*, we have Kansuke's own reference to the use of the Tanegashima-teppō and the widespread proliferation of these matchlocks in the sixth decade of the sixteenth century. This great heihosha, himself, was killed by shot at Kawanakajima. The present study relies on Obata Toshihiro's translation published by Hawley Publications, California, 1994, under the title *Heihō Okugishō*.
20 Griffith, *Sun Tzu*, VI v.6. Ts'ao Ts'ao, (155–220), King of Wei.
21 Miyamoto Musashi was once criticized for writing about heihō when he was about fifty because if he had waited a few more years before committing his thoughts to paper 'he would have gained more valuable experience' (insight). Such a comment is still to be found in modern Kendo circles.
22 *Sun Tzu* III, 1: 'Generally in war the best policy is to take a state intact; to ruin it is inferior to this.'

NOTES

Chapter 2: Who Were the Bugeisha?

1 For details of Yamamoto Kansuke's life, refer to Knutsen (2004), *Rediscovering Budo*, and Obata Toshihiro, *Heihō Okugishō*.
2 Sawyer, *The Seven Military Classics of Ancient China*, p. 17.
3 *Sun Tzu*, Book V. 12.
4 For a fuller discussion of yang and yin refer to D.C. Lau and Roger Ames, *Sun Pin , The Art of Warfare*, p. 103–108.
5 T'ang T'ai-tsung usurped the throne in 627 CE. It is thought that this work may be a record of conversations between Li Ching (571–649) and the Emperor.
6 The observations here are based on the forms demonstrated by the Omote-dō kata in the Kashima Shinto-ryū-no-heihō.

Chapter 3: Ch'i Within the Eishin-ryū

1 Kendō *ku-dan Hanshi*. He was one of the last successors of the Kurama-ryū Iai-jutsu. 'It is quite reasonable to suppose that some forms of combative drawing from the scabbard in direct attack existed from the Heian period but no actual records survive.'
2 There are a number of variant forms of the names employed by masters of the Eishin-ryū; for the purposes of my text I have used those names, rendered in Romaji, taught me more than forty years ago in West Japan. For others, refer to Shimabukuro and Pellman, *Flashing Steel*. I should add that my master received his instruction from the Eighteenth Headmaster, Masaoka Kazumi, before the Second World War.
3 *Kai-waza*: A form or technique deriving from another, an alternative development.
4 See discussion by Gen. Sam Griffith, *Sun Tzu, The Art of War*.
5 Roald Knutsen, *Rediscovering Budo*, pp. 212–20.
6 *Uji-gami*: A tutelary deity; an ancestral line of descent.
7 *Nihon Kendō-shi*, (1919).
8 *Reimu*: A prophetic vision; an inspired dream; a divine revelation in a dream. (Kenkyusha dictionary definition.)
9 *Bugei*: the Art of War, to study the techniques of the battlefield.
10 *Tachi-ai*: Walking forms.
11 See David Hall, *Marishiten, etc.* and Knutsen, *Rediscovering Budo*.
12 Refer to *Rediscovering Budo* for a brief discussion of the reality of 'Ninjutsu'.

Chapter 4: Hasegawa Eishin-ryū Structure

1 Kamō Jisaku, 1898–1980.
2 Such a change within the Jigen-ryū resulted in the offshoot of the tradition, the Yakugan Jigen-ryū, being suspended within the Shimazu-han for quite a long period in the eighteenth century, according to Yamada Jirokichi, *Nihon Kendo-shi*, and Kuroki Toshihiro (in personal discussion, Saga-shi, April 1976).

Chapter 5: Mao Tse-tung and Unorthodox Tactics

1 Mao Tse-tung, *On Guerrilla Warfare*, translated and introduced by Samuel B. Griffith, 1978.
2 *On Guerrilla Warfare*, p. 21.
3 *On Guerrilla Warfare*, p. 23.

4 *On Guerrilla Warfare*, p. 22.
5 *On Guerrilla Warfare*, p. 27.
6 *Tameshigiri*: testing the quality of a blade or a technique by cutting a dummy 'body'. In modern times, a dummy made of tightly bound straw; in former times, the body of a condemned criminal and sometimes several.
7 In 1930, Mao Tse-tung used a similar paraphrase of Sun Tzu in *A Single Spark can start a Prairie Fire*, p. 70. In his letter from the Front Committee to the Central Committee concerning guerrilla tactics, he wrote: 'The enemy advances, we retreat; the enemy camps, we harass; the enemy tires, we attack; the enemy retreats, we pursue.'
8 *Sun Tzu*, Book VI, v. 5.

Chapter 6: Iai-jutsu Seen as Flexible Warfare

1 Mao Tsu-tung, *Strategy in China's Revolutionary War* (December 1936) p. 141.
2 *On Protracted War*, p. 245.
3 *Sun Tzu*: Book VII, 15.
4 A.L. Sadler expressed this view in his *History of Japan* and it has been quoted a number of times since; however, I am inclined to think this may not have been the case.
5 For more details see David Hall, *Marishiten etc.*, pp. 250–1, and Knutsen, *Rediscovering Budo*, pp. 190–1.
6 Oral transmission = *denshō-bugaku*.
7 During the Edo period and continued to the present day, many if not all, the ko-ryu awarded their advancing students a denshō known as a *menkyo* that defined their level of proficiency; the fully initiated deshi was awarded the highest level denshō called *menkyō-kaiden*. This highly honourable licence permitted teaching at the discretion of the recipient. At the present time it is not known when this system was introduced.
8 After Lau and Ames. Gen. Griffith translates this passage as: 'He whose generals are able and not interfered with by the sovereign will be victorious' (III, 20).
9 See *Rediscovering Budo*, pp. 59–61. For a fuller account of Yamamoto Kansuke's early life.
10 Lau and Ames, *Sun Pin, etc.*, p. 54.
11 *Sun Tzu*, V. 2.
12 Takeda Nobutora had planned to disinherit Harunobu (Shingen) in favour of his younger brother, Nobuyoshi. This caused the rebellion.
13 See *Rediscovering Budo* for more details of this master's life.
14 The *Art of War*, III, 19–22.
15 Takeda Katsuyori and his son, Nobukatsu, committed *seppuku* after the battle of Temmoku-zan in Kai province in 1582. It spelt the effective end of the Takeda-han, the succession passing north to a cadet branch, the Nambu-han.
16 The *Art of War*, III, 31.

Chapter 7: Foreknowledge

1 Simplicity or directness as opposed to floridness and clutter is often to be noted when one observes classical Iai-jutsu kata and contrasts with some systems of Iai-dō developed in the mid- to late-Edo period or after.
2 Yoshikawa Koichirō *sensei*, Headmaster, Kashima Shintō-ryū, in personal teaching, March 1976.

NOTES

3. Lau and Ames, *Sun Pin, etc.*, p. 75.
4. Three *'fechtbuch'* were produced by Talhoffer for the private instruction of his patron.
5. *Sun Tzu*, V.9.
6. Readers are referred to *Rediscovering Budo* for a more detailed discussion of this subject.
7. Refer to Knutsen and Knutsen, *Japanese Spears*, for comments on the postures used in So-jutsu and Naginata-jutsu in general.
8. Lau and Ames, *Sun Pin*, etc., p. 85.
9. *Sun Tzu*, XI, 31.
10. There is a possibility that the *kongō-no-kamae* may have originated in the Chujō-ryū but this is not certain.
11. Gen.Tao Hanzhang, (trans.), *Sun Tzu's Art of War*, 1987, p. 61.
12. Yamasaki, *Shingon, etc.*, pp. 85–9.
13. E. Dale Saunders, *Mudrā, etc.*, p. 185.
14. *Rediscovering Budo*, pp. 231–8.
15. *Seigan-no-kamae*: 'aiming at the eye' posture.

Chapter 8: The Distinction in the Heihō between *Ch'i* and *Cheng*

1. *On Guerrilla Warfare*, Ch. II, p. 48.
2. *On Guerrilla Warfare*, p. 50.
3. *Heihō-Okigishō*, trans. Obata Toshihiro, p. 118.
4. Some expert opinion, actually voiced by Kuroki Toshihiro, expresses doubt that Miyamoto Musashi was the actual author of the *Gorin-no-shō*. (Oh, what heresy!).

Chapter 9: The Influence of the Mountain Religion

1. See *Japanese Spears*.
2. Yamada Jirokichi, David Hall, and Knutsen.
3. Yamamoto Kansuke, *Heihō-Okigishō*.
4. Refer to Richard Hayes, 'Palaeolithic Adaptive Traits and the Fighting Man', in *Hoplos*, 4. 2 (June 1984).
5. See *Rediscovering Budo*, pp. 26–7.
6. *Sanrei*: enlightenment bestowed by the deity of a mountain.
7. Symbolized by the three-legged crow.

Chapter 10: The Esoteric Principles Contained in *In* and *Yō*

1. E. Dale Saunders, *Mudrā*, p. 117.
2. Saunders further notes that in Tibet, Mārīcī was regarded as the goddess of the sunrise, and in Japan she is supposed to reside in one of the stars of the Great Bear (Ursa Major). Ibid. p. 117.
3. Sun Tzu, V. 6 and 7.
4. Kuroki Toshihiro suggested that the Puyŏ-Kayan attempt to settle in southeastern Kyushu may not have come from tribes sailing from Korea but actually moving west from Kashima before their final migration eastwards again through the Inland Sea.
5. Some authorities identify these images as 'dragon-like' but their arguments for this seem weak.

NOTES

6. Otake Risuke, Katori-Shintō-ryū, in personal conversation with the author and the late Donn Draeger, Narita-shi, January 1976.
7. See David Hall, *Marishiten*, etc.
8. See *Rediscovering Budo*.
9. Michael Gorman, personal conversation, Okayama-ken, October 2002.
10. Knutsen and Knutsen, *Japanese Spears*.
11. H. Russell Robinson, in personal discussion, November 1957.
12. *Japanese Spears*, p. 74.
13. *Japanese Spears*, pp. 16–17 and Figs 18 and 19.
14. *Gosannen Kassen Emaki*: this work exists in three scrolls today but was thought to have originally been composed in four as the account begins at the halfway stage. It is the work of Korehisa Hida-no-kami in the Kamakura period, fourteenth century. In the *Ping-fa*, (9, 22), Sun Tzu wrote: 'Birds rising in flight is a sign that the enemy is lying in ambush; when the wild animals are startled and flee he is trying to take you unaware.'

Chapter 11: Unexpected Attacks Against an Unprepared Enemy

1. '*Atsumori*' written by Seami Motokiyo (1363–1444).
2. Griffiths, *Sun Tzu, The Art of War*, p. 134.
3. There are a number of accounts in English describing this battle. (See Sadler, Dening and Sansom).
4. If this decline was noticeable just two generations removed from the Sengoku period, then just how much further has understanding sunk in the three centuries that have passed *since* then?
5. *Sun Tzu*: VI, 5, and VII, 12. (Trans. by Samuel Griffiths.)
6. *SunTzu*, III, 28.
7. Okada Morihiro *Hanshi*, in personal instruction, Tokyo, April 1970, endorsed many times by Kamō Jisaku in Kyushu, 1975–80.
8. *Sun Tzu*: III, 10.
9. *Sun Tzu*: VII, 27.
10. Although in most circumstances this situation is a relatively low-level security problem, the great danger is the feeling of complacency that guard detachments often feel when not actually faced by an obvious threat. Several incidents following the Second Gulf War illustrate the results only too clearly.
11. This Iai form even makes provision for an 'orderly' retreat.
12. *Tsuka-koshira*: metal pommel mounting at the end of the hilt. Often used as a useful extra weapon by swordsmen.

Chapter 13: Warfare and Ritual

1. Yamamoto Kansuke, writing in his *Heihō-Okigishō* sometime before 1561 brushed the following: 'Throughout the year there are superstitions. However, in strategy, let your enemy be the one to be deeply influenced by the superstition, not you.' (Trans. Obata Toshihiro.)
2. The fifteenth century German *fechtmeister*, Hans Talhoffer, went into great detail on this subject in his first manuscript of 1443, indicating that the prevailing beliefs at that time in Central and Western Europe held that judicial battle and probably more general warfare was governed by either St Mary or St George.
3. One suggested link is set out in *Rediscovering Budo*, p. 197.

NOTES

4   See Donn F. Draeger, *Classical Bujutsu* for more detailed information.
5   *Heihō-Okigishō*.
6   *Rediscovering Budo*.
7   Tower of London, Ms. 1–33.
8   Martin Wierschin, *Meister Johann Liechtenauers Kunst des Fechtens*, pp. 67–75.
9   Nuremberg, Germanisches Museum, 30 Cod. Ms. 3227a.
10 Wierschin, p. 71.
11 Translation by my father, William Pedre Erland Knutsen.
12 Wierschin, p. 71.
13 Sigmund Ringeck, Dresden Sachsische Landesbibliothek, Ms. C. 487 (Single: D), first half of the fifteenth century. Ringeck could have been retained by Albrecht I, of Niederbaiern-Straubing, 1358–1404. A precise dating of this manuscript is not possible.
14 I. The Gotha Ms. Chart. A 558 of 1443/48. II. The Ambraser Codex No. 55 of 1459. III. The third is the Codex. Icon. 394a of 1467.
15 The *Flos Duellatorum* was published in facsimile in 1902 under the title *Il Fior di Battaglia*.
16 Hans Burgkmaier, *Die Weisskünig*, No. 37. (Fol. 176b, Cod. 3933, Vienna).
17 It is a curious matter that while all warfare and combat has the objective of attaining victory and the Swiss and German techniques were brutally direct, there is an element of spiteful menace in these Italian forms and especially in the accompanying verses.
18 See *Rediscovering Budo* for further comment on this subject.
19 The term 'therianthropic' derives from two Greek words: *therion*, meaning 'wild beast', and *anthropos*, meaning 'man'. It is used to describe the many different beast-man figures that are common throughout prehistory and folklore in every part of the world. Some authorities consider they are the oldest and deepest rooted of all man's creative imaginings.
20 *Migi-hara-tombo*: A one-handed posture with the sword held at the right hip, left foot leading. *Katate-kasumi-jōdan*: Sword held one-handed above the head with the blade tip pointing forwards as at the opponent, in this case with the left hand held out, palm outwards as though warning the adversary, left foot advanced.
21 The noses of the earliest of these 'tengu masks' used in the Bugaku dances that date from around the eighth and ninth century are usually show a rather sausage shaped proboscis and never the 'Roman' nose form found in the martial scrolls.
22 See *Rediscovering Budo*.
23 See *The Life-Giving Sword* by Yagyu Muneyoshi, trans. by William S. Wilson, Kodansha International Ltd, Tokyo, 2003.
24 These denshō were examined by the author at the Department of History at Saga Daigakuin in March and April 1976, and quite detailed discussions held with the late Dr Kuroki Toshihiro, head of the Department at that time. Dr Kuroki had specialized in the background influence of the mikkyō on the bugei and had focused in particular on the Shinkage-ryū, the Jigen-ryū, and the Taisha-ryū, as these very old entities were strong in the Kyushu region.
25 It should be noted that where *suneate* or *kyahan* are worn in both these ryū, they appear in all identifiable cases to be correctly secured with their *himo* (cords or ties) at the front of the shins.

Chapter 14: Conclusions

1. These quotations from *Sun Tzu* are taken from Book VI, 22, 23, and 25.
2. Araki Matayemon was an excellent swordsman of the Nitō-ryū, founded by Miyamoto Musashi about ten years previously. The katahi-uchi at Sagara Igano-kuni took place about 1630. This kenshi is buried in the Genchu-ji temple, Tottori-shi, and his grave is very well kept. An account of the this famous fight is to be found in Mitford's *Tales of Old Japan* under 'Kazuma's Revenge' but there are several other traditions extant.
3. *Heihō Okigishō*, p. 94.

Chapter 15: 'Subtle and Insubstantial . . .'

1. All 37,000 people who took to the shelter of the re-fortified Hara Castle on the island of Amakusa were put to the sword; men, women and children. Many Japanese historians would have us believe that this was a Christian revolt, and some Christians were amongst the defenders, it is true, but the true reason was the excessive cruelty of Matsukura Shigeharu, the local lord, and Terazawa Katataka, another daimyō. See Sansom, *A History of Japan,* Vol. III, pp. 37 and 38.

Glossary of Terms

BAKUFU   A military government in the medieval period, e.g. The Muromachi Bakufu, etc.
BAKUMATSU   The disturbed period between 1856 and 1877 during the Restoration of the Emperor Meiji.
BŌ-JUTSU   The art of using a long staff, (six shaku or more in length). In some very old ryū such a staff was regarded as a yari that had lost its blade. Bō-jutsu is believed to have been developed by the early yamabushi and their skills subsequently studied by the buke.
BOKKEN-SHOBŪ   A contest using wooden swords.
BUGEI   The Arts of the Battlefield.
BUGEISHA   One who practises the military arts.
BUKE   Military families tracing their ancestry back to the Imperial line or ancient chieftains.
BUNKAI   Analysis or discussion, definitions of a technique.

CHENG   In Chinese, orthodox strategies.
CH'I   In Chinese, unorthodox trategies.
CHI-KEN-IN   The mudrā of the Knowledge Fist; the Sword-mudrā, thought to bestow power over enemies.
CHŌTŌ   Thought to be a Muromachi period term for a long-hilted sword or a short-shafted halberd.

GLOSSARY OF TERMS

DAI   In a title, meaning 'great' or 'upper', e.g. Dai-Sendatsu: a senior initiate in Shugendō, or Daimyō, a 'great-name' i.e. a great lord.
DESHI   A disciple or student.
DENSHŌ   A scroll recording teaching forms.

EMAKIMONO   A picture scroll. (See FUZEIKEN below.)
EMBU   A display of the military arts, usually within the grounds of a shrine.

FUZEIKEN   Artistic; having the meaning of 'embellished' or 'illustrated' when describing those tora-no-maki containing painted figures of tengu-swordsmen, such as in the tengu scrolls.

GE   A crow-billed halberd blade, usually bronze, current in the Yayoi period, 300 BCE – 300 CE.
GENPEI WAR   Struggle between the Minamoto and the Taira clans, 1180–85.
GEKKEN   Name given to early Kendō styles in the late-Tokugawa period.
GUNBAI-UCHIWA   A war-fan usually carried by Heihō-sha, often inscribed with astrological tables and/or the symbols of Marishi-ten.

HAN   In the medieval period, a clan group.
HANMI   A 'Half-turned' position frequently encountered in Bujutsu. Such a posture presents a reduced target for an enemy.
HANSHI   Foremost teaching degree, especially in modern Kendō and Iai.
HARA-TOMBO   Sword postures where the sword is held horizontally at waist height threatening a sharp forwards thrust.
HEIHŌ-JIN   A strategist of middle rank.
HEIHŌ-SHA   A strategist of the highest ability.
HEIHŌ-ZUKAI   A strategist of low rank.
HONZON   A visual aid to prayer; an image of a deity.
HSUNG-NU   Ancient nation of nomadic tribes living along the northern and western borders of China, esp. the Altai and the Gobi regions.

IAI-HIZA   A half-kneeling posture used by swordsmen in Iai-jutsu.
IAI-JUTSU   The warrior Art of Drawing-Sword, not to be confused with the later development into IAI-DŌ.
IN   That which is obscured; unorthodox or secret. Chinese: YIN
IN-NO-KAMAE   A 'shaded' posture often adopted by swordsmen.
INOSHISHI   A wild boar, often employed as a symbol of Marishi-ten in the mikkyō.

JIBARI   An initial probing feint attack used in some very early ko-ryū traditions of the fifteenth and sixteenth centuries.

KA   An ancient form of crow-billed halberd in China. A type also found across the whole Euro-Asian landmass. (See GE.)
KAGE   'Shaded', 'Hidden', 'Secret'.
KAI-WAZA   An alternative technique or form.
KASHIRA   The pommel mount at the end of a tsuba, or sword hilt.
KASUMI   'Mist', 'Hazy', 'Obscured'.
KATAHI-UCHI   A 'vengeance' fight; a custom often employed by warriors for reasons of maintaining honour and so on. Increasingly usurped by commoners in the late-Edo period.
KENSHI   A expert swordsman.
KESAGIRI   An oblique sword or halberd stroke delivered downwards through either shoulder to hip or upwards from hip to shoulder.
KIRIOROSHI   In Iai-jutsu, the final powerful downwards cut, the *coup-de-grâce*. (Also termed NUKITSUKE.)
KIRI-TE   A sharp sideways flick of the sword intended to 'shake off blood' from the blade after use but to leave the blade ready for instant use.
KO-RYŪ   Any named transmitted style originating before the Edo period.
KUMITACHI   A term often used to define private practice in ko bujutsu. An alternative word is 'Kumidachi'.
KUMIUCHI   The battlefield art of grappling with or without armour. Within the bugei there were three distinct levels of kumi-uchi defined.
KURASU-TOBI   'Crow-hopping or jumping'. (See TENGU-TOBI.)

GLOSSARY OF TERMS

LIU-T'AO   Name of a Chinese treatise on warfare extant around 1000 CE.

MA-JUTSU   Sleight of hand, trickery purporting to be magic.
MARISHI-TEN (MĀRĪSHĪ)   Female war deity whose ancient cult, in various forms, spread across the whole Euro-Asian landmass.
MARISHITEN-HOBYŌ-IN   (See ONGYŌ-IN.)
MENKYO-KAIDEN   Licence confirming full proficiency and understanding of an art, especially in the Bugei (Arts of War). Possession of menkyo-kaiden is considered a great honour.
MIKKYŌ   Esoteric rituals and teachings in Buddhism and Shugendō.
MON   A clan or family device worn as a badge. Not heraldic within the European meaning.
MUDRĀ   Wide range of hand postures – 'seals' – employed in the Buddhist and Shugendō mikkyō, either with or without accompanying prayer.
MUSHA-SHUGYŌ   To travel for severe training, especially in the bugei.
MUSHA-SHUGYŌSHA   One who undergoes such severe training in order to gain understanding.

NUKITSUKE   The second or final cut following the draw cut in Iai. (See KIRIOROSHI.)

OKUDEN   A 'secret' technique or teaching principle.
ŌMOTE-DŌ   Usually the introductory or lowest level taught within a ryū; an 'Outer-Gateway' series of forms.
ONGYŌ-HO   Sorcery, a trick, an illusion.
ONGYŌ-IN   The mudrā of 'Hiding Forms', particularly valued by the followers of Marishi-ten.
Ō-TORA-NO-MAKI   A 'Great Tiger' scroll; usually a secret scroll recording the outlines of a transmission in the bugei.

PUYŎ   The name of the nomadic tribal group that swept down through Korea in ancient times, part of whom seized lands in Japan after first century CE. This latter group known as the PUYŎ-

180

GLOSSARY OF TERMS

KAYAN. By the fifth century these tribes were known as the YAMATO.

REIMU A sudden moment of enlightenment.
RYŪ A transmitted system of teaching principles.

SANKAKU-SHINKŌ The 'folk religions or beliefs' of the indiginous inhabitants of Japan. A modern term describing the 'Mountain Religion'.
SAN-LUEH A Chinese text on warfare extant around 1000 CE.
SENDATSU An initiate, esp. in Shugendō.
SENGOKU PERIOD The 'Age of the Country at War', ca. 1470–1575.
SHAKU Old Japanese measure of length roughly equal to 11⅞; inches.
SHIDACHI In swordsmanship, one who studies how to oppose an attack in practice.
SHIH (Chinese) Foreknowledge.
SHINGYAKU A religious novice; someone who aspires to initiation.
SHINKEN-SHOBU A match using 'live' blades.
SHINOBU To 'steal-in', to infiltrate.
SHISHI A Chinese lion.
SHISHI-FUNJIN A violent attack like that of a lion.
SHŌMEN A vertical cut with a sword delivered at the head.
SHUGENDŌ An esoteric sect, partly Buddhist and partly 'Sankaku-shinkō'.
SŌ-JUTSU The battlefield art of spearmanship.
SONSHI Japanese form of Sun Tzu. Often used to mean Sun Tzu's *Art of War*.
SUKASHI-TSUBA A swordguard, usually made of iron, with a pierced design.

TACHI A slung sword worn on the left hip.
TACHI-AI Sword forms made from standing postures.
TAMESHIGIRI Techniques of testing the cutting power of a sword-blade; to 'prove' a blade, usually on condemned criminals or, latterly, on a straw target.

TA-RYŪ Forms in the bugei that do not relate to the teachings of any classical tradition; their purpose is to prevent the leaking of secret information.

TA-RYŪ-JIAI Matches between bugeisha of different ryū.

TATE-HIZA A half-kneeling sitting posture adopted by classical warriors. (See IAI-HIZA.)

TE-BOKO Odd-shaped 'halberds' preserved in the Shōsō-in, Nara, dating from the seventh century CE.

TENGU-MICHI A popular alternative name given to the 'Yamabushi-michi' or paths. (See YAMABUSHI-MICHI.)

TENGU-SHŌ Transmissions of secret principles by means of a tengu. An upper-level in a bugei tradition.

TENGU-TOBI A distinctive and sudden jump often found in the KATA of some very old traditions in the bugei. Also found in Shugendō ritual. Sometimes called KURASU-TOBI. Some ko-bujutsu experts teach that the proper time to employ this type of jumping attack is at dusk or twilight in order to confuse the enemy.

TENGU-TOBIKIRI-NO-JUTSU The art of delivering a TENGU-TOBI attack.

TESSEN A fan.

TORA-NO-MAKI A secret teaching scroll. (See Ō-TORA-NO-MAKI.)

TSUBA The swordguard that protects the hand of a swordsman. Mostly made of metal. Sometimes present on Naginata. (See SUKASHI-TSUBA.)

TSUBA-ZERIAI In swordsmanship, the entanglement of the sword guards in close-space action. This was always a dangerous position, particularly for the less-experienced combatant.

TSUKA The hilt of a Japanese sword.

UCHIDACHI In swordsmanship and the spear arts, one who 'attacks' in formal practice. (See SHIDACHI.)

UJI-GAMI Ancestors, ancestral spirits; the progenitors of a group.

UMA-JIRUSHI Often thought of as 'a horse-leading samurai', a close retainer of comparatively high rank. In the late-Heian period the term 'uma-jirushi' meant an act of expressing obeisance to one's master. Source: *Azuma-kagami* (History of the East) dating from the early-thirteenth century. The more modern interpretation appears to date from the late-Muromachi period according to several Kendō masters of high rank.

WAZA   A 'form', in Bujutsu and Budō.

YAMABUSHI   An ascetic priest, especially one belonging to the Shugendō sect.
YAMABUSHI-MICHI   The 'Way of the Yamabushi', i.e. Shugendō.
YANG   Chinese; that which is in the light; orthodox. Yō in Japanese.
YATAGARASU   A 'three-legged crow'; symbol of an ancient mystery and sect associated with the Kumano group of Shintō shrines.
YIN   Chinese; that which is obscured; unorthodox or secret. Japanese: IN.
YO-NO-KAMAE   A 'clear' posture, usually with the sword held at the left shoulder.
YUBI-NO-MAKURA   A sword or spear movement that protects the exposed shoulder and back whilst in the process of striking back. Sometimes known as the 'Pillow-Dream' posture amongst swordsmen.

ZANSHIN   Awareness, 'remaining spirit'.

Bibliography

Alhouse Green, Miranda and Stephen, *The Quest for the Shaman. Shape-shifters, Sorcerers and and Spirit-healers of Ancient Europe*, London: Thames & Hudson, 2005.

Barfield, Thomas J., *The Perilous Frontier: nomadic empires and China*, Blackwell, 1989.

Barnes, Gina L., *State Formation in Korea: Historical and Archaeological Perspectives*, Curzon Press, 2001.

—— *The Rise of Civilization. The Archaeology of China, Korea and Japan*, London: Thames and Hudson, Pbk, 1999.

Bekink, Willem, *Erfenis van den goden*, Rijswijk: Elmar B.V., 2003.

Blacker, Carmen, *The Catalpa Bow*, London: George Allen & Unwin, 1975.

British Kendo Renmei, *Hasegawa Eishin-ryū Iai Hyōhō*, 1982.

Bunker, Emma C., *Ancient Bronzes of the Eastern Eurasian Steppes*, Arthur M. New York: Sackler Foundation, 1997.

Clottes, Jean, and Lewis-Williams, D., *The Shamans of Prehistory: Trance and Magic in the Painted Caves*, New York: Harry N. Abrams, 1998.

Confucius, *The Anelects*, Trans: D. C. Lau, Penguin Classics, 1979.

—— *The Anelects*, Trans: Arthur Waley, Allen & Unwin, 1938.

Cosmo, Nicola di, *Ancient China and Its Enemies: The Rise of Nomadic Power in East Asian History*. Cambridge U.P., 2002.

Dann, Jeffrey Lewis, *Poetry in the Martial Ways*, Unpub. Provisional text, Mito, 1975.

Draeger, Donn F., *Classical Bujutsu*, The Martial Arts and Ways of Japan, Vol. 1, New York and Tokyo: Weatherhill, 1973.

—— *Classical Budo*, The Martial Arts and Ways of Japan, Vol. 2, New York and Tokyo: Weatherhill, 1973.

—— *Modern Bujutsu and Budo*, The Martial Arts and Ways of Japan, Vol. 3, New York and Tokyo: Weatherhill, 1974.

Funazaki Takashi, 'Sohei, the Fighting Monks of Medieval Japan' in *The East*, Vol. 36, No. 6, March/April 2001.

Gorai Shigeru, *Yama no Shūkyō* (Mountain Religion) (Japanese language). Tokyo: Kudokawa-shoten, 1991.

BIBLIOGRAPHY

Gorman, Michael S.F., *The Quest for Kibi and the True Origins of Japan*, Bangkok, 1999.
Guilaine, J. and Zammit, J., *The Origins of War: Violence in Prehistory*, (Translated by Melanie Hersey), Blackwell Pub., 2005.
Hancock, Graham, *Supernatural: Meetings with the Ancient Teachers of Mankind*, Century, 2005.
Hall, David Avalon, *Marishiten, Buddhism and the Warrior Goddess*, Unpub. Ph.D. Thesis, Univ. of California, Berkeley, 1967.
Hori Ichio, *Folk Religion in Japan*, Chicago UP., 1958.
Iida Heizaemon, *Kashima-shi*, (History of Kashima), (Japanese language).
Jimbutsu Orai-sha, *Japanese War Scroll Paintings*, Tokyo, 1962, (Japanese text with English summary).
Keeley, Lawrence H., *War Before Civilisation, The Myth of the Peaceful Savage*. OUP., 1994.
Knutsen, Roald M., *Japanese Polearms*, London: The Holland Press, 1963.
——*Rediscovering Budo, From a Swordsman's Perspective*, Folkestone: Global Oriental, 2004.
Knutsen, Roald & Patricia, *Japanese Spears, Polearms and their Use in Old Japan*, Folkestone: Global Oriental, 2004.
Kuroki Toshihiro, *Historical View of Bujutsu Origins*, (Japanese Language), Saga, Japan, ca. 1975.
——*Hyōhōkadenshō (of Shinkage-ryū)*, (Japanese language with brief English notes), Saga, Japan.
Ledyard, Gari, *Galloping Along with the Horseriders: Looking for the Founders of Japan*, Columbia University, 1974.
Lewis-Williams, D., *The Mind in the Cave*, London: Thames & Hudson, 2002.
Mao Tse-tung, *On Guerrilla Warfare*, Trans. Samuel B. Griffith, USA: Anchor Books,1978.
——*Selected Military Writings*, Beijing: Foreign Languages Press, 1963.
Masaoka Kazumi, 'Eishin-ryū Iai Hyōhō' in *Budō Hyuron*, (Japanese language Journal), 1967.
Masayuki Shimabukuro and Leonard J. Pellman, *Flashing Steel, Mastering Eishin-ryū Swordsmanship*, Berkeley, California: Frog Ltd, 1995.
Metropolitan Mus. of Art, *The Golden Deer of Eurasia. Treasures from the Russian Steppes*, New York: Met. Mus. of Art, 2000.
McNeilly, Mark, *Sun Tzu and the Art of Modern Warfare*, OUP, 2001.
Miyake Hitoshi, *The Mandala of the Mountain: Shugendō and Folk Religion*, Keio Ed. Gaynor Sakamori UP, 2005.
Miyamoto Kesao, *Tengu to Shugenja*, (Tengu and Shugenja) (Japanese language), Kyoto: Jinbun-shoin, 1992.
Murayama Shuichi, *Shugen no Sekai*, (The World of Shugendō) (Japanese language), Kyoto: Jinbun-shoin, 1992.
Otake Risuke, *The Deity and the Sword: Katori Shinto-ryu*, Trans. Donn F. Draeger, 3 Vols, Tokyo: Minato Research and Publishing, 1977.
Papinot, E., *Historical and Geographical Dictionary of Japan*, Vermont & Tokyo: Tuttle, 1980.
Reeder, Ellen D. (Ed.), *Scythian Gold, Treasures from the Ancient Ukraine*, New York: Harry N. Abrams Inc., 1999.
Robinson, H.R., *Japanese Arms and Armour*, London, 1969.

BIBLIOGRAPHY

Rudenko S.I., *Frozen Tombs of Siberia: The Pazyryk Burials of Iron Age Horsemen*, Trans. M.W.Thompson, London, 1970.
——*Koultura Gornogo Altaiä*, Catalogue Art russe, Paris, 1967.
Sansom, Sir George, *A History of Japan*, 3 Vols, London, 1958.
Sasamori Junzo, *Ittō-ryū Gokui*, (Japanese language), Tokyo.
Sato Hiroaki, *Legends of the Samurai*, New York: Overlook Press, 1995.
Saunders, E. Dale, *Mudrá;*, Bolingen Series, Princeton UP, 1960.
Sawa Tadaaki, *Art in Japanese Esoteric Buddhism*, Tokyo: Weatherhill, 1976.
Sawyer, Ralph D., *The Seven Military Classics of Ancient China*, Oxford: Westview Press, 1993.
Sinclaire, Clive, *Samurai, The weapons and spirit of the Japanese warrior*, London: Salamander, 2001.
Skoss, Diane (Ed.), *Koryu Bujutsu, Classical Warrior Traditions of Japan*, New Jersey: Koryu Books, 1997.
Sugawara Makoto, *The Ancient Samurai*, Tokyo: The East, 1986.
——*Lives of Master Swordsmen*, Tokyo: The East, 1985.
Sun Pin, *Sun Pin, The Art of Warfare*, Trans. and commentary by Lau, D.C. and Ames, R., New York: Ballantine Books, 1996.
Sun Tzu, *The Art of War*, Trans. Samuel B. Griffith, Oxford UP, 1963.
Takano Hiromasa, *Heihō Ittō-ryu*, (Japanese language), Tokyo, 1985.
Tao Hanzhang, *Sun Tzu's Art of War*, New York: Sterling Publishing, 2000.
Visser, Marius W. de , 'The Tengu', in *Transactions of the Asiatic Society of Japan*, Vol. XXXVI, 1908.
Watanabe Tadashige, *Shinkage-ryu Sword Techniques*, Vol. I, Tokyo, 1993.
Watatani Kiyoshi, *Nihon Kendō Hyakusen*, (Japanese language), Tokyo: Akita Shoten, 1972.
——*Bugei Ryūha Daijiten*, (Japanese language), Tokyo, 1978.
Wierschin, Martin, *Meister Johann Liechtenauers Kunst des Fechtens*, München, 1965.
Wontak Hong, *Paekche of Korea and the Origin of Yamato Japan*, Kudara International, 1994.
Yamada Jirokichi, *Nihon Kendū-shi*, (Japanese Language), Tokyo, 1925.
——*Kashima Shinden Jikishinkage-ryu*, (Japanese language), Tokyo, 1927.
Yamamoto Kansuke, *Heihū Okigishū, The Secret of High Strategy*, Trans. Obata Toshihiro, Hollywood, California, 1994.

Index

Aisu Ikkyō (master swordsman), 10, 149, 152.
Aizu-han, 51
Amaterasu (Sun deity), 18, 19
Amida Buddha, 20
Araki Matayemon (master swordsman), 163

Bakufu (military government) definition, 177
Bakumatsu, 166, 177
Battō ('Drawing-Sword Arts'), 6, 36, 87, 166
Bō, Bō-jutsu, 17, 108–109, 154, 177
Bugei, definition, 177
Bu-no-jutsu, Bu-no-ri, 96–7

Chang-Yü (commentator on Sun-Tzu), 113
Chia Lin (Chinese commentator), 7
Chi-ken-in, mudrā, 86, 177, Plate 28
Chōtō, definition, (also Nagatsuka-no-katana), 177
Chujō-ryū-no-heihō, 84–5

Dainichi-kyō, 85
Dainichi Nyōrai, 85
Dengaku-hazama, (battle), 113
Denshō (scrolls), 33, 80–2, 147, 153, 178

Emaki-mono, or Gunki Emaki-mono, 75, 109, 111,178

Flags and Banners, (manipulation of), 31
'Four Pillars' of Bujutsu, 26, 110
'Four Poisons' of the Bugei, 6, 34, 117–19, 137
Fudō Myō-ō, 21, 99–100, 104, 134, 154, 162, Plate 28
Futsu-nushi-no-kami (Katori deity), 19, 101, 103
Fuzeiken (definition), 178

Genpei War, 178
Gosennen Kassen Emaki, 111
Guerrilla strategy, definition, 88, 119–21, 133
Gunbai-uchiwa (war fan), 136, 178
Gunki Emaki-mono (see Emaki-mono)

Hara-tombo sword posture definition), 178
Hasegawa Eishin-ryū, 4, 36–45, 46–51, 51, 69, 115, 125, 167
Hayashizaki Jinsuke (master swordsman), 40–2, 44, 122–3, 125, 126, 128, 130, 132, 152, 167
Hayashizaki Shigenobu-ryū, 40–1
Heihō Hidenshō, 24

INDEX

Heihō-jin, 29, 178
Heihō Kadenshō, 102
Heihō Okigishō, 109, 162, 164
Heihō-sha, 29, 133, 178
Heihō-zukai, 29, 178
Hostage system (to ensure good faith), 67–8
Honzon (votive images), 178
Hsung-nu (nomadic tribes), 84, 105, 108, 178
Hugin and Munin ('messengers' of Odin), 22

I-ching (Book of Changes), 82
Ichi-no-tani, (battle), 112
Idzumo region, 18, 105
In-no-kamae (sword posture), 82–3, 179
Inoshishi (wild boar symbolism), 179, Plate 29

Jibari (initial probing with sword), 179
Jingō-kōgō, (semi-mythical chieftainess), 18

Kage-no-ryū, 149
Kamiidzumi Nobutsuna (master swordsman), founder of Shinkage-ryū, 10, 149, 152, Plate 4
Kanemaki-ryū (developed from Chūjō-ryū), 84
Kashima, 18, 99, 105, 161
Kashima Shintō-ryū, Frontispiece, 26, 75, 81, 164, Plates 5, 7, 19, 21, 25–7
Katori, 18, 19, 21, 106
Katori Shintō-ryū, also Tenshin-Shōden Katori-Shintō-ryū, 81, Plate 24
Ketsu-miko-no-kami (deity), 20
Kitabatake Tomonori (powerful warrior-aristocrat), 67
Kōbō Daishi (founder of Shingon-shū), 85
Kofun period, 7
Kojiki (ancient chronicle), 18
Kumano sanzan (shrines) and region, 18, 19, 21, 106

Kyō-ryū, 136
Kyū-jutsu, 110

Liechtenauer, Johann (German master swordsman), 137 *et seq.*, 148–9
Liu Bocheng (PLA general), 9
Liu-t'ao (Chinese military treatise), 62, 180

Marume Kurandō-dayū (master swordsman), founder of Taisha-ryū, 149, 152, Plate 1
Mao Tse-tung, 5, 13, 52–8, 60, 83, 88–9, 91, 93, 114, 123, 133
Matsumoto Bizen-no-kami (master swordsman), 164
Marishi-ten (Mārīcī), 7, 9, 20–1, 39, 41, 65, 85, 87, 95–6, 99–100, 102–103, 107, 110, 125, 134–6, 148, 153–4, 156, 162, 180, Plate 29
Mencius, 2
Minamoto-no-Yoshi-ie, 111
Minamoto-no-Yoshitsune, 62, 99, 112–14, 155
Mikkyō (esoteric religious practices), 10, 21, 85, 93, 95, 106, 154, 157, 161, 180
Miyamoto Musashi (master swordsman), 11, 92–3, 96
Mudrā (esoteric hand 'seals'), 180
Munin ('messenger' of Odin) (see Hugin), 22
Musha-shugyō (severe training), 11, 180
Musō Jikiden Eishin-ryū (Hasegawa Eishin-ryū, 48
Musō Kenshin-ryū (*denshō*), 153

Nagashino (battle), 68
Nagatsuka-no-katana (Chōtō), 79
Naginata (glaive), Naginata-jutsu, 17, 36, 78–9, 109–110, 154, 166, Plates 19, 20
Nihongi (ancient chronicle), 18
Nyumon-shiki (induction ceremony), 14, 63, 65

INDEX

Odin (Norse deity), 22
Oe Tadafusa (bugei expert), 111
Okuden (secret transmissions in the Bugei), 7, 39, 180
O-kuni-nushi (line of ancient rulers, semi-mythical), 18, 105
Ōmine-okugake-michi, or Kumano-kōdō (yamabushi paths), 21
Ōmori-ryū Iai, 42, 46, 69, 115, 167
On Guerrilla Warfare (Ch. *Yu Chi Chan*), 53, 58
Ōnō-ha Ittō-ryū, 83–5
Ō-tora-no-maki (transmission scrolls), 180

Peoples Liberation Army (PLA), 9
Premeriacco, Fiore da. (Italian weapon master), 142–5
Puyŏ, or Puyŏ-Kayan, 16, 17, 84, 95, 99, 101–102, 105, 108, 135, 148, 180

Ramilles, (battle) (1706), 32
Reimu (sudden intuitive understanding), 181
Rodriguez, John (Jesuit priest, 16c.), 158

Sankaku-shinkō (mountain (folk) beliefs), 16–17, 98, 106, 181
San Lueh (Chinese military treatise), 62, 181
Satori (enlightenment in mikkyŏ), 11
Secret agents, 6, 71–2
Sekigahara (battle), 4, 14, 168
'*Seven Military Classics*' (of China), 3, 12, 15, 28, 30, 61
'Shaping' the enemy, 6, 31–3, 56, 131, 133, 165
Shimabara Rebellion, 168
Shimazu-han (clan), 51
Shingan-ryū, 148, Plate 28
Shingon-shū, 10, 44, 84–5, 100
Shinobi (or Shinobu), (infiltration), 44, 181
Shiten-ryū Kiome-no-waza, 86
Shugendō, 10, 15, 19, 21, 44, 85, 93, 100–106, 161, 181

Sō-jutsu (spearmanship), 36, 108–10, 166, 181
Sungari River, 15
Sun Pin (Chinese military philosopher), 2, 61, 63, 75, 95, 156

Taisha-ryū, (founded by Marume Kurandō-dayu), 149, 151–3, Plate 29
Takeda-han (clan), 15, 28, 66–9
Takeda Katsuyori (son of Takeda Shingen), 68
Take-mika-dzouchi-no-kami (deity of Kashima Shrine), 18–19, 101, 103
Talhoffer, Hans (German master-at-arms), 77, 79–82, 136–7
Tally Fleet (to China), 148
Tao, Taoism, 8, 17, 44, 81, 84–6, 95, 106, 122
Tao Hanzhang (PLAP general), 9, 84
Tendai-shū, 8
Tengu ('messengers' of Marishi-ten), 8, 18, 22–3, 39, 41, 81, 100, 110, 121, 124, 134, 146–55, 182, Plates 10, 11, 28
Tengu Hidenshō (scroll), 153
Tenshin Shōden Katori Shintō-ryū, 154, Plate 15
Toda Seigen (16c. general and master swordsman), 64–5
Tokugawa Bakufu, 5
Tora-no-maki (transmission scrolls), 19, 33, 76 *et seq.*, 100, 182
Tosa-han (clan), 51
Tottori region, 18
Ts'ao Ts'ao (Chinese military philosopher), 25
Tsukahara Bokuden (master strategist and swordsman), 10, 26, 55, 67, 70

Udō Myō-jin Shrine, 162

Vajra (trident symbol in the mikkyŏ), 85

Warrior-shaman (Kofun period), 8

INDEX

Warring States period (China), 30, 84, 108

Yagyū Shingan-ryū, Plate 16
Yagyū Shinkage-ryū, 102, 149–51, Plates 2, 3, 17, 18, 22, 23
Yamabushi, 8–10, 17, 21, 96, 106, 110, 148, 162, 183
Yamamoto Kansuke, (master strategist), 15, 24–6, 28, 65–7, 70, 92, 96, 109, 136, 162–4, 166, 168
Yamato, 17, 20
Yang and Yin; definitions, 32
Yatagarasu, 17–20, 22, 183, Plate 29
Yayoi people, 8, 16, 20
Yu Chi Chan (see *On Guerrilla Warfare*), 53, 58

Zanshin (awareness), 59–60, 86, 97, 183
Zao-dō (Yoshinoyama), 21
Zao-gongen (fierce deity), 100, 104
Zen Buddhism, 10, 23, 100